Intro to Archaeology & Geology

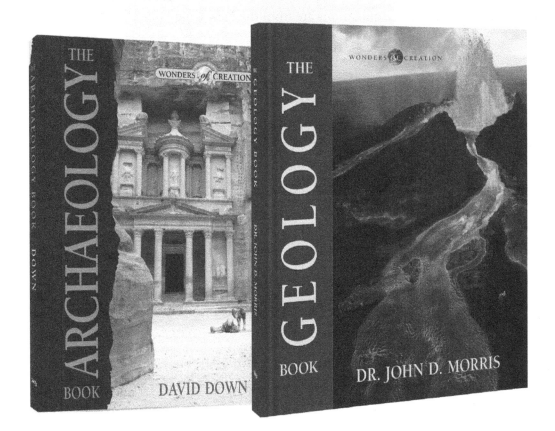

LESSON PLANNER

- Weekly Lesson Schedule
- Student Worksheets
- Quizzes & Test
- Answer Key

7th – 9th grade

1 Year Science

1/2 Credit

First printing: March 2013
Second printing: July 2013

Master Books®, P.O. Box 726, Green Forest, AR 72638

Master Books® is a division of the New Leaf Publishing Group, Inc.

ISBN: 978-0-89051-728-4

Unless otherwise noted, Scripture quotations are from the New King James Version of the Bible.

Printed in the United States of America

Please visit our website for other great titles:

www.masterbooks.net

For information regarding author interviews,
please contact the publicity department at (870) 438-5288

Master
Books®
A Division of New Leaf Publishing Group
www.masterbooks.net

Where Creation Inspires Education

Since 1975, Master Books has been providing educational resources based on a biblical worldview to students of all ages. At the heart of these resources is our firm belief in a literal six-day creation, a young earth, the global Flood as revealed in Genesis 1–11, and other vital evidence to help build a critical foundation of scriptural authority for everyone. By equipping students with biblical truths and their key connection to the world of science and history, it is our hope they will be able to defend their faith in a skeptical, fallen world.

If the foundations are destroyed, what can the righteous do?
Psalm 11:3; NKJV

As the largest publisher of creation science materials in the world, Master Books is honored to partner with our authors and educators, including:

Ken Ham of Answers in Genesis
Dr. John Morris and Dr. Jason Lisle of the Institute for Creation Research
Dr. Donald DeYoung and Michael Oard of the Creation Research Society
Dr. James Stobaugh, John Hudson Tiner, Rick and Marilyn Boyer, Dr. Tom Derosa, and so many more!

Whether a pre-school learner or a scholar seeking an advanced degree, we offer a wonderful selection of award-winning resources for all ages and educational levels.

But sanctify the Lord God in your hearts, and always be ready
to give a defense to everyone who asks you a reason for the hope
that is in you, with meekness and fear.
1 Peter 3:15; NKJV

Permission to Copy

Lessons for a 36-week course!

Overview: This *Introduction to Archaeology and Geology PLP* contains materials for use with *The Archaeology Book* and *The Geology Book* in the Wonders of Creation series. Materials are organized by book in the following sections:

📇	Study Guide Worksheets
Q	Quizzes
T	Semester Tests
🔑	Answer Key

Suggested Optional Science Lab
See page 12

Features: Each suggested weekly schedule has three easy-to-manage lessons which combine reading, worksheets, and vocabulary-building opportunities including an expanded glossary for each book. Designed to allow your student to be independent, materials in this resource are divided by section so you can remove quizzes, tests, and answer keys before beginning the coursework. As always, you are encouraged to adjust the schedule and materials needed to in order to best work within your educational program.

Workflow: Students will read the pages in their book and then complete each section of the PLP. They should be encouraged to complete as many of the activities and projects as possible as well. Tests are given at regular intervals with space to record each grade. If used with younger students, they may be given the option of only choosing activities or projects of interest to them and taking open book tests.

Lesson Scheduling: Space is given for assignment dates. There is flexibility in scheduling. For example, the parent may opt for a M–W schedule rather than a M, W, F schedule. Each week listed has five days but due to vacations the school work week may not be M–F. Adapt the days to your school schedule. As the student completes each assignment, he/she should put an "X" in the box.

🕐	Approximately 30 to 45 minutes per lesson, two to three days a week
🔑	Includes answer keys for worksheets, quizzes, and semester tests
📇	Worksheets for each chapter
🔄	Quizzes are included to help reinforce learning and provide assessment opportunities; optional semester exams included
📄	Designed for grades 7 to 9 in a one-year course to earn 1/2 science credit

Course includesbooks from creationist authors with solid, biblical worldviews:

David Down - *The Archaeology Book*
David Down has been a field archaeologist for over four decades, excavating regularly in Israel and involved in numerous digs over the years.

Dr. John Morris - *The Geology Book*
Dr. John Morris is president of the Institute for Creation Research. He received his Doctorate in Geological Engineering at the University of Oklahoma. He held the position of Professor of Geology before being appointed President in 1996. He currently travels and speaks on the topic of creation science.

Contents

Introduction to Archaeology and Geology

Course Description

This is the suggested course sequence that allows one core area of science to be studied per semester. You can change the sequence of the semesters per the needs or interests of your student; materials for each semester are independent of one another to allow flexibility.

Semester 1: Archaeology

The Archaeology Book takes you on an exciting exploration of history and ancient cultures. You will learn both the techniques of the archaeologist and the accounts of some of the richest discoveries of the Middle East that demonstrate the accuracy and historicity of the Bible. You will unearth: how archaeologists know what life was like in the past; why broken pottery can tell more than gold or treasure; some of the difficulties in dating ancient artifacts; how the brilliance of ancient cultures demonstrates God's creation; history of ancient cultures, including the Hittites, Babylonians, and Egyptians; the early development of the alphabet and its impact on discovery; the numerous archaeological finds that confirm biblical history; and why the Dead Sea scrolls are considered such a vital breakthrough. Filled with vivid full-color photos, detailed drawings and maps, you will have access to some of the greatest biblical mysteries ever uncovered.

Semester 2: Geology

Rocks firmly anchored to the ground and rocks floating through space fascinate us. Jewelry, houses, and roads are just some of the ways we use what has been made from geologic processes to advance civilization. Whether scrambling over a rocky beach or gazing at spectacular meteor showers, we can't get enough of geology! *The Geology Book* will teach: what really carved the Grand Canyon; how thick the Earth's crust is; why the Earth is unique for life; the varied features of the Earth's surface from plains to peaks; how sedimentary deposition occurs through water, wind, and ice; effects of erosion; ways in which sediments become sedimentary rock; fossilization and the age of the dinosaurs; the powerful effects of volcanic activity; continental drift theory; radioisotope and carbon dating; and geologic processes of the past. Our planet is a most suitable home. Its practical benefits are also enhanced by the sheer beauty of rolling hills, solitary plains, churning seas and rivers, and majestic mountains—all set in place by processes that are relevant to today's entire poplulation of this spinning rock we call home.

First Semester Suggested Daily Schedule

Date	Day	Assignment	Due Date	✓	Grade
		First Semester-First Quarter — **The Archaeology Book**			
Week 1	Day 1	Read Pages 6-14 • The Archaeology Book • (AB)			
	Day 2				
	Day 3	Read Pages 15-19 • (AB)			
	Day 4				
	Day 5	What Archaeology is . . . - Terms to Know • **Archaeology Ch1: Worksheet 1** • Pages 15-16 • Lesson Plan • (LP)			
Week 2	Day 6	What Archaeology is All About - Questions **Archaeology Ch1: Worksheet 1** • Page 16 • (LP)			
	Day 7				
	Day 8	What Archaeology is All About - Activities **Archaeology Ch1: Worksheet 1** • Page 16 • (LP)			
	Day 9				
	Day 10	Read Pages 20-29 • (AB)			
Week 3	Day 11				
	Day 12	Land of Egypt - Terms to Know **Archaeology Ch2: Worksheet 1** • Page 17 • (LP)			
	Day 13				
	Day 14	Land of Egypt - Questions **Archaeology Ch2: Worksheet 1** • Pages 17-18 • (LP)			
	Day 15				
Week 4	Day 16	Land of Egypt - Activities **Archaeology Ch2: Worksheet 1** • Page 18 • (LP)			
	Day 17				
	Day 18	Chapter 1-2 Study Day			
	Day 19				
	Day 20	Chapter 1-2 Quiz 1 • Page 61 • (LP)			
Week 5	Day 21				
	Day 22	Read Pages 30-35 • (AB)			
	Day 23				
	Day 24	The Hittites - Terms to Know, Questions **Archaeology Ch3: Worksheet 1** • Pages 19-20 • (LP)			
	Day 25				
Week 6	Day 26				
	Day 27	The Hittites - Activities **Archaeology Ch3: Worksheet 1** • Page 20 • (LP)			
	Day 28				
	Day 29	Read Pages 36-41 • (AB)			
	Day 30				

Date	Day	Assignment	Due Date	✓	Grade
Week 7	Day 31	Ur...Chaldees - Terms to Know, Questions **Archaeology Ch4: Worksheet 1** • Pages 21-22 • (LP)			
	Day 32				
	Day 33	Ur of the Chaldees - Activities **Archaeology Ch4: Worksheet 1** • Page 22 • (LP)			
	Day 34				
	Day 35	Read Pages 42-45 • (AB)			
Week 8	Day 36				
	Day 37	Assyria - Terms to Know, Questions **Archaeology Ch5: Worksheet 1** • Pages 23-24 • (LP)			
	Day 38				
	Day 39	Assyria - Activities **Archaeology Ch5: Worksheet 1** • Page 24 • (LP)			
	Day 40				
Week 9	Day 41	The Archaeology Book Chapters 1-5 Study Day			
	Day 42				
	Day 43	**The Archaeology Book Ch1-5 Quiz 2** • Page 63 • (LP)			
	Day 44				
	Day 45	Read Pages 46-51 • (AB)			
First Semester-Second Quarter					
Week 1	Day 46	Babylon:...Gold - Terms to Know, Questions **Archaeology Ch6: Worksheet 1** • Pages 25-26 • (LP)			
	Day 47				
	Day 48	Babylon: City of Gold - Activities **Archaeology Ch6: Worksheet 1** • Page 26 • (LP)			
	Day 49				
	Day 50	Read Pages 52-59 • (AB)			
Week 2	Day 51	Persia - Terms to Know, Questions **Archaeology Ch7: Worksheet 1** • Pages 27-28 • (LP)			
	Day 52				
	Day 53	Persia - Activities **Archaeology Ch7: Worksheet 1** • Page 28 • (LP)			
	Day 54				
	Day 55	Read Pages 60-64 • (AB)			
Week 3	Day 56				
	Day 57	Read Pages 65-69 • (AB)			
	Day 58				
	Day 59	Petra - Terms to Know, Questions **Archaeology Ch8: Worksheet 1** • Pages 29-30 • (LP)			
	Day 60				

Date	Day	Assignment	Due Date	✓	Grade
Week 4	Day 61	Petra - Activities **Archaeology Ch8: Worksheet 1** • Page 30 • (LP)			
	Day 62				
	Day 63	Chapter 6-8 Study Day			
	Day 64				
	Day 65	Chapter 6-8 Quiz 3 • Page 65 • (LP)			
Week 5	Day 66	Read Pages 70-77 • (AB)			
	Day 67				
	Day 68	The Phoenicians - Terms to Know, Questions **Archaeology Ch9: Worksheet 1** • Pages 31-32 • (LP)			
	Day 69				
	Day 70	The Phoenicians - Activities **Archaeology Ch9: Worksheet 1** • Page 32 • (LP)			
Week 6	Day 71	Read Pages 78-83 • (AB)			
	Day 72				
	Day 73	Dead Sea Scrolls - Terms to Know, Questions **Archaeology Ch10: Worksheet 1** • Pages 33-34 • (LP)			
	Day 74				
	Day 75	Dead Sea Scrolls - Activities **Archaeology Ch10: Worksheet 1** • Page 34 • (LP)			
Week 7	Day 76				
	Day 77	Read Pages 84-93 • (AB)			
	Day 78				
	Day 79	Israel - Terms to Know, Questions **Archaeology Ch11: Worksheet 1** • Pages 35-36 • (LP)			
	Day 80				
Week 8	Day 81				
	Day 82	Israel - Activities **Archaeology Ch11: Worksheet 1** • Page 36 • (LP)			
	Day 83				
	Day 84	The Archaeology Book Chapters 6-11 Study Day			
	Day 85				
Week 9	Day 86				
	Day 87	The Archaeology Book Ch 6-11 Quiz 4 • Page 67 • (LP)			
	Day 88				
	Day 89	The Archaeology Book Chapters 1-11 Study Day			
	Day 90				
		Test (optional) • Page 69 • (LP)			
		Mid-Term Grade			

Second Semester Suggested Daily Schedule

Date	Day	Assignment	✓	Grade
		Second Semester-Third Quarter — **The Geology Book**		
	Day 91			
	Day 92	Read Pages 4-9 • The Geology Book • (GB) Planet Earth - Words to Know, Questions **Geology Intro. & Ch1: Worksheet 1** • Pages 39-40 • (LP)		
Week 1	Day 93			
	Day 94	Planet Earth - Activities **Geology Intro. & Ch1: Worksheet 1** • Page 40 • (LP)		
	Day 95			
	Day 96	Read Pages 10-19 • (GB)		
	Day 97			
Week 2	Day 98	The Ground We Stand On - Words to Know, Questions **Geology Ch2: Worksheet 1** • Pages 41-42 • (LP)		
	Day 99			
	Day 100	The Ground We Stand On - Activities **Geology Ch2: Worksheet 1** • Page 42 • (LP)		
	Day 101	Read Pages 20-27 • (GB)		
	Day 102			
Week 3	Day 103	The Earth's Surface - Words to Know **Geology Ch3: Worksheet 1** • Page 43 • (LP)		
	Day 104			
	Day 105	The Earth's Surface - Questions **Geology Ch3: Worksheet 1** • Pagse 43-44 • (LP)		
	Day 106	The Earth's Surface - Activities **Geology Ch3: Worksheet 1** • Page 44 • (LP)		
	Day 107			
Week 4	Day 108	The Geology Book Chapters 1-3 Study Day		
	Day 109			
	Day 110	The Geology Book Chapters 1-3 Quiz 1 • Page 73 • (LP)		
	Day 111	Read Pages 28-35 • (GB)		
	Day 112			
Week 5	Day 113	Geological Processes and Rates - Words to Know **Geology Ch4: Worksheet 1** • Page 45 • (LP)		
	Day 114			
	Day 115	Geological Processes and Rates - Questions **Geology Ch4: Worksheet 1** • Pages 45-46 • (LP)		

Date	Day	Assignment	✓	Grade
Week 6	Day 116			
	Day 117	Geological Processes and Rates - Activities **Geology Ch4: Worksheet 1** • Page 46 • (LP)		
	Day 118			
	Day 119	Read Pages 36-41 • (GB)		
	Day 120			
Week 7	Day 121	Geological Processes and Rates - Words to Know **Geology Ch4: Worksheet 2** • Page 47 • (LP)		
	Day 122			
	Day 123	Geological Processes and Rates - Questions **Geology Ch4: Worksheet 2** • Pages 47-48 • (LP)		
	Day 124			
	Day 125	Geological Processes and Rates - Activities **Geology Ch4: Worksheet 2** • Page 48 • (LP)		
Week 8	Day 126	Read Pages 42-48 • (GB)		
	Day 127			
	Day 128	Geological Processes and Rates - Words to Know **Geology Ch4: Worksheet 3** • Page 49 • (LP)		
	Day 129			
	Day 130	Geological Processes and Rates - Questions **Geology Ch4: Worksheet 3** • Pages 49-50 • (LP)		
Week 9	Day 131			
	Day 132	Geological Processes and Rates - Activities **Geology Ch4: Worksheet 3** • Page 50 • (LP)		
	Day 133			
	Day 134	Read Pages 48-53 • (GB)		
	Day 135			
Second Semester-Fourth Quarter				
Week 1	Day 136			
	Day 137	Geological Processes and Rates - Words to Know **Geology Ch4: Worksheet 4** • Page 51 • (LP)		
	Day 138			
	Day 139	Geological Processes and Rates - Questions **Geology Ch4: Worksheet 4** • Pages 51-52 • (LP)		
	Day 140			
Week 2	Day 141			
	Day 142	The Geology Book Chapter 4 Study Day		
	Day 143			
	Day 144	The Geology Book Chapter 4 Quiz 2 • Page 75 • (LP)		
	Day 145			

Date	Day	Assignment	✓	Grade
Week 3	Day 146	Read Pages 54-57 • (GB)		
	Day 147			
	Day 148	Ways to Date the Entire Earth - Words to Know **Geology Ch5: Worksheet 1** • Page 53 • (LP)		
	Day 149			
	Day 150	Ways to Date the Entire Earth - Questions **Geology Ch5: Worksheet 1** • Pages 53-54 • (LP)		
Week 4	Day 151	Ways to Date the Entire Earth - Activities **Geology Ch5: Worksheet 1** • Page 54 • (LP)		
	Day 152			
	Day 153	Read Pages 58-68 • (GB)		
	Day 154			
	Day 155	Great Geologic Events of the Past - Words to Know **Geology Ch6: Worksheet 1** • Page 55 • (LP)		
Week 5	Day 156	Great Geologic Events of the Past - Questions **Geology Ch6: Worksheet 1** • Pages 55-56 • (LP)		
	Day 157			
	Day 158	Great Geologic Events of the Past - Activities **Geology Ch6: Worksheet 1** • Page 56 • (LP)		
	Day 159			
	Day 160	The Geology Book Chapters 5-6 Study Day		
Week 6	Day 161	The Geology Book Chapters 5-6 Quiz 3 • Page 77 • (LP)		
	Day 162			
	Day 163	Read Pages 69-72 • (GB)		
	Day 164			
	Day 165	Questions People Ask - Words to Know **Geology Ch7: Worksheet 1** • Page 57 • (LP)		
Week 7	Day 166			
	Day 167	Questions People Ask - Questions **Geology Ch7: Worksheet 1** • Pages 57-58 • (LP)		
	Day 168			
	Day 169	Read Pages 73-75 • (GB)		
	Day 170			
Week 8	Day 171			
	Day 172	The Geology Book Chapters 7-8 Study Day		
	Day 173			
	Day 174	The Geology Book Chapters 7-8 Quiz 4 • Page 79 • (LP)		
	Day 175			

Date	Day	Assignment	✓	Grade
Week 9	Day 176	The Geology Book Chapters 1-8 Study Day		
	Day 177			
	Day 178	The Geology Book Test • Page 81 • (LP) optional		
	Day 179			
	Day 180			
		Semester Exam		
		Final Grade		

Suggested Optional Science Lab

There are a variety of companies that offer science labs that complement our courses. These items are only suggestions, not requirements, and they are not included in the daily schedule. We have tried to find materials that are free of evolutionary teaching, but please review any materials you may purchase. The following items are available from www.HomeTrainingTools.com.

Intro to Archaeology & Geology
RM-GEOBAG Geology Field Trip in a Bag
RM-ROCKMIN Rocks & Minerals of the U.S. Basic Set

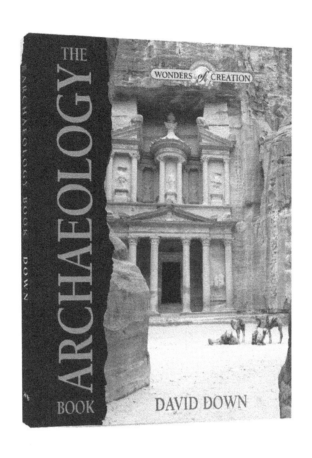

Archaeology Worksheets for use with

The Archaeology Book

Name

Words to Know – Write the definition in the space provided below.

accession year

AD

archaeology

artifact

BC

carbon dating

ceramic

chronology

debris

EB

exile

exodus

hieroglyphs

LB

MB

millennium

non-accession year

pottery

synchronism

tell

Questions

1. What does the word *archaeology* mean?

2. For what three reasons were cities built on hills?

3. When did people first start using coins?

4. Why are inscriptions found on ancient pottery valuable to archaeologists?

5. What are the four main periods of archaeological time?

Activities

1. Find a small piece of damp clay or plasticine, and with the end of a screwdriver impress your name on it, creating your own seal impression.

2. Set up an archaeological treasure hunt with some everyday items. Have an adult bury the items in shallow holes, covering them with a thin layer of soil. Carefully go about digging them up and classifying your treasures in a journal.

Words to Know – Write the definition in the space provided below.

Asiatic

baulk

dowry

drachma

dynasty

mastabas

Nubia

Pharaoh

Questions

1. What is the Egyptian name for Egypt?

2. Who was the first Egyptian king to build a pyramid?

3. Who built the biggest pyramid in Egypt?

4. What was the name of the Egyptian god of the Nile River?

5. What did the Egyptians in the twelfth dynasty mix with their bricks to hold them together?

Activities

1. Find a small cardboard or plastic box. Make some mud out of earth and a little water, mix some dry grass with it and put it into the box. When it is fairly dry turn the box upside down and lift it off the brick you have made. Let it dry.

2. Develop a chart with your family history or dynasty. Try to trace the ancestry of one parent or both, depending on the information you have available. List these as names on a graph or draw an actual tree with the branches representing the family members.

Words to Know – Write the definition in the space provided below.

amphitheater

Anatolia

bathhouse

inscription

Questions

1. Which was the strongest nation in the Middle East 3000 years ago?

2. Which two nations did the Syrians think had come to attack them?

3. Who were the Hittites descended from?

4. How often were the Hittites mentioned in the King James Version of the Bible?

5. Who wrote the book *The Empire of the Hittites*?

Activities

1. Draw a rough map of Turkey and write in the names Constantinople and Boghazkale where you think they were a few thousands years ago.

2. Read the book of Esther. Write three discussion questions about the story and find an evening this week to discuss the story with your family.

Name _____

Words to Know – Write the definition in the space provided below.

centurion

Chaldees

nomad

papyrus

Questions

1. In the Bible, how many references are there to Ur of the Chaldees?

2. Who was the main excavator of Ur of the Chaldees?

3. Why did Woolley not excavate the cemetery as soon as he found it?

4. What was the name of the people who occupied ancient Ur?

5. What did Woolley find in the Death Pits of Ur?

Activity

1. Research the ancient Hittite civilization online or at your local library. How many resources can you find available for this people once thought to be a myth?

Words to Know – Write the definition in the space provided below.

bulla

Medes

scarab

seal

Questions

1. Who discovered Nineveh?

2. What was the name of the ruins where Layard first started digging?

3. What was the name of the king of Israel that was mentioned on the black pillar Layard found in Nimrud?

4. What was the name of the king of Israel when Sennacherib besieged Jerusalem?

5. How many cities did Sennacherib claim he conquered?

Activities

1. Get some plasticine or soft clay and press it down flat. Get a pencil or small twig of a tree the width of a pencil, and cut one end to form a triangle. Press this end down horizontally and vertically on the clay, making your own cuneiform impressions.

2. Sometimes archaeologists must learn a language to help them understand a culture better, just as Layard studied the Persian language. Choose a language to study briefly and obtain several books from the library to help you learn some basic words and phrases.

Name

Words to Know – Write the definition in the space provided below.

Armenians

cuneiform

strata

syncline

Questions

1. What was the name of the cuneiform record which told a story similar to the Bible record of Noah and the flood?

2. Which Assyrian king compiled a library of tablets in Nineveh?

3. What did the Babel builders stick their bricks together with?

4. Which king made Babylon a city of gold?

5. Which Bible prophet predicted that Babylon would become uninhabited?

Activities

1. Log on to the Internet and search for "Ishtar Gate Berlin Museum." This will bring up a picture of the gate from Babylon that Professor Koldewey sent back to Berlin.

2. Research the Bible account of the Flood and compare this account with other accounts from around the world. A good resource from a Christian perspective is *Flood Legends* by Charles Martin.

Words to Know – Write the definition in the space provided below.

Persia

rhyton

Questions

1. Who was the king who first carved out the Medo-Persian Empire?

2. In what year did he conquer Babylon?

3. Which Persian king left an inscription on the rock face of the Zagros Mountain near Bisitun?

4. What was the name of the great Persian city that Darius built?

5. What was the name of the official who tried to destroy all the Jews in Persia?

Activities

1. Read the book of Esther in the Bible and count how many times the word God is used. You may be surprised.

2. Study the celebration of Purim that is still celebrated today. Observe how the traditions and even the games relate back to Queen Esther.

Name

Words to Know – Write the definition in the space provided below.

bedouin

cistern

Edom

Edomites

Nabataeans

siq

theater

wadi

Questions

1. In what year did Burckhardt discover Petra?

2. Whose descendants occupied Petra?

3. What were his descendants called?

4. Which Bible prophet wrote a book about Petra?

5. Which Roman emperor had a road constructed through Petra?

Activities

1. Pottery in Petra was very thin. Get some plasticine or clay and make a small teacup without a handle. See how thin you can make it.

2. Set up a tent in your yard and talk about what it would be like to live life as a Bedouin, wandering from place to place. Consider staying overnight in the tent, weather permitting.

Words to Know – Write the definition in the space provided below.

Baal

causeway

Yehovah

Questions

1. What were the four main cities of ancient Phoenicia?

2. Which trees were Phoenicia famous for?

3. Whose tomb did Pierre Montet find?

4. Which Bible prophet challenged the prophets of Baal?

5. Which Bible prophet predicted that ancient Tyre would never be found?

Activities

1. Find a map of the Mediterranean Sea and try to work out how far it is from Phoenicia—now called Lebanon, to Spain. That is how far Phoenician ships sailed.

2. Make a relief of your hand by pressing your palm and fingers into a flat piece of clay or by pressing foil over your hand to make a metallic looking imprint. See how much detail you can add once the initial impression is made.

Name

Words to Know – Write the definition in the space provided below.

scroll

vellum

Questions

1. In what year was the first Dead Sea Scroll found?

2. How many letters were in the Hebrew alphabet?

3. What were most of the Dead Sea Scrolls written on?

4. Which is the longest acrostic in the Bible?

5. What was the name of the settlement near the cave where the Dead Sea Scrolls were found?

Activities

1. Find a King James version of the Bible and look at Psalm 119. At the beginning of every eighth verse you will find a letter of the Hebrew alphabet. Write out the 22 letters of the alphabet.

2. Take several pieces of white or tan cardstock. Write or paint a favorite Bible passage across it. When dry, roll up the "scroll" and tie it off with ribbon or string.

Name _____

Words to Know – Write the definition in the space provided below.

annunciation

Calvary

Golgotha

grotto

Messiah

ossuary

Passover

Questions

1. Which Roman Emperor adopted Christianity as the state religion?

2. Jesus' name in Hebrew was Yeshua. What does that mean?

3. In what river was Jesus baptized?

4. In which city did Jesus enter a synagogue and cast out a demon?

5. What does the word *Calvary* mean?

Activities

1. At the back of most Bibles are maps of Palestine. Try to calculate how far it was from Jerusalem to Galilee. Jesus walked this distance many times.

2. Using poster board or cardstock, create a map of Israel. You might consider paints or markers to color the rivers, lakes, and land. Glue on small blocks or other objects to represent towns and cities.

Geology Worksheets for use with

The Geology Book

Name

Scripture: Genesis 1:1–31; Genesis 3:17–21; Romans 6:23; Romans 8:22

Words to Know – Write the definition in the space provided below.

Principle of uniformity

Principle of catastrophe

Asthenosphere

Plate

Questions

1. Operational science is the science that deals with repeatable, observable experiments in the present. Origins science deals with reconstructing events that have happened in the past. What is the key difference between "origins" and "operational" science?

2. There are two ways of thinking about the unobserved past. What are they?

3. Where is the true history of the earth found?

4. In what order did God create the heavens and earth? (e.g., describe what He created on Day 1, Day 2, etc.) See Genesis 1.

5. Write a short paragraph answering the question, "What is sin?"

6. What are the main "zones" into which the earth is divided?

7. What is the earth's crust composed of?

8. What is the purpose of the earth's atmosphere?

Activities

Review the text on pages 4-10. Two views of earth history are compared (uniformity and catastrophe). Make a chart of the comparisons: see if you can find 3 to 5 examples to include in your comparison.

Name _____

Scripture: Genesis 1:1; Obadiah 1:3

Words to Know – Write the definition in the space provided below.

Igneous rocks

Sedimentary rocks

Metamorphic rocks

Ripple marks

Crossbed

Concretions

Metamorphism

Questions

1. This chapter lists three categories of rock, with each category containing a discussion on several types of rock. Draw an expanded version of the table on the next page.

   a. In the first column, list each type of rock mentioned in this chapter.

   b. In the second column, list the category under which the rock is found.

   c. In the third column, describe the composition of each rock type.

   d. In the fourth column, describe how the rock is formed.

   e. In the fifth column, make a list of where the rock is found today.

   f. Watch out for types within types! (We've done the first one for you!)

Type	Catagory	Composition	Formation	Found
Granite	Igneous	Quartz and feldspar with mica and hornblende	Formed when molten rock is cooled	Mountains Upper mantle

Activities

Start collecting stones/small rocks from around your area (or other areas to which you travel). Try to classify the type of rock you have found. Can you find samples of each rock you described in the above table?

Note: If you go to a National/State/local park, ask permission to remove the stones/rocks you are collecting. Do not remove any rocks or stones from someone's garden without permission.

Name _____

Scripture: Genesis 8:4; Psalm 121

Words to Know – Write the definition in the space provided below.

Plain

Sediment

Plateau

Mountain

Canyon

Questions

1. Why are low-lying plains considered good for farmland?

2. Three types of plateaus are mentioned. Write a short description of each type and include examples for each.

3. Four types of mountains are mentioned. Write a short description of each type and include examples for each.

4. What process causes the formation of mesas and buttes?

Activities

This "experiment" will take a few weeks. Erect a mound of dirt in your backyard (the pile of dirt should be at least three feet high). Visit the mound each day and record the following information: height of mound, width of mound. You will notice that the mound will get shorter and the base wider. Determine what could have caused the difference in height and width. Was it the wind? Was it rain? Was it a dry-spell? Have fun with this.

Name

Scripture: Genesis 8

Words to Know – Write the definition in the space provided below.

Erosion

Deposition

Turbidite

Questions

1. List the five primary causes of normal erosion.

2. List and describe the three rapid erosive processes explained in this chapter.

3. How is a turbidite formed?

4. List two types of events that could cause deposition to happen quickly and on a large scale.

Activities

Observe and record the types of erosion you find near where you live. If possible, take pictures to document your findings.

Name

Scripture: Genesis 7, 8

Words to Know – Write the definition in the space provided below.

Compaction

Cementation

Fossils

Petrification

Gastrolith

Coprolite

Questions

1. Sand is made primarily of what mineral?

2. Write a paragraph describing how sedimentary rocks are formed.

3. What is the main condition required for a fossil to form?

4. There are many different types of fossils. Name at least four.

5. In you own words, write a description of how dinosaur fossils were formed.

Activities

NOTE: This week's activities reinforce the fact that fossils and petrified wood can be made over a short period of time.

- Make a "fossil" using plaster of paris. Mix the plaster of paris and pour it onto a paper plate. Gather different objects (leaves, toy dinosaurs, etc.) and press each into the plaster of paris. Lift the object off and include a print of your hand or foot. Let dry. Paint when dry, if desired.

- This is week 3 of the experiment started in Lesson 3. Has your mound decreased any? What caused this decrease?

Name

Scripture: Genesis 7, 8

Words to Know – Write the definition in the space provided below.

Volcanism

Fumaroles

Geyser

Fault

Earthquake

Questions

1. At least two ways that volcanoes erupt are discussed in the text. What are they?

2. When forces in the earth's crust build to a breaking point along a fault, the sections move in one of three ways. What are they? Write a short description of each type.

3. What factors influence whether a rock will "bend" or "break"?

4. Write two paragraphs explaining the arguments given in the book for and against continental separation.

Activities

1. You might want to do a more in-depth study of volcanoes. Research some of the more famous volcanoes that have erupted in the past. When did they erupt? How devastating to the surrounding areas were they? What types of devastation did they cause? How often do these volcanoes erupt? How long do the eruptions last?

2. This is week 4 of the experiment started in Lesson 3. Has your mound decreased any? What caused this decrease?

Name

Scripture: Psalm 18

Words to Know – Write the definition in the space provided below.

Atom

Isotope

Radioisotope dating

Carbon dating

Questions

1. In your own words, describe the different ways metamorphic rocks are thought to form.

2. What are unstable atoms called? What is the most well-known radioactive atom?

3. How are the unstable qualities of uranium useful to mankind?

4. What is the difference between a parent isotope and a daughter isotope?

5. Explain the process of carbon dating in determining when a plant died.

Name

Scripture: Psalm 18

Word to Know – Write the definition in the space provided below.

Magnetic field

Uplift

Questions

1. List some of the methods currently being used to determine the age of the earth.

2. What must we consider when evaluating conclusions obtained from these methods?

3. What conclusion can be drawn about the age of the earth from the various dating methods discussed in this chapter?

NOTE: Be sure to read the picture captions!

Activities

This is a simple experiment to study wind erosion. Stick a piece of two-sided tape on one side of several stirring paddles (you can get these from a paint store) and place them in the ground in various spots around your yard. Make sure they don't all face in one direction; have one face north, another south, etc. At regular intervals check the amount of dust or soil sticking to the tape. Depending on the amount of wind and the direction from which it blows, you will see that some paddles collect more dust than the other paddles. More soil will stick to the paddles where wind erosion is taking place. What is in the way that is preventing dirt from sticking to those paddles where little is collected?

Name

Scripture: Genesis 1–11; Romans 6:23

Words to Know – Write the definition in the space provided below.

Second law of science

Fountains of the deep

Glacier

Polar ice cap

Questions

1. What four events have had the greatest impact in shaping the earth's geology?

2. How did the creation event affect the earth's geology?

3. What role has the Fall played in shaping the earth?

4. What was the cause of the global Flood? What were some of the geological results of the Flood?

5. Why do we find ocean fossils near the top of Mt. Everest?

6. Draw a map of North America. Outline the extent of the ice covering at the peak of the Ice Age (hint: be familiar with the map on page 67).

Activities

Write a two-page comparative essay on the cause of the Ice Age. Compare the views presented in *The Geology Book* with secular views on this subject. Your essay should include an introduction, your thesis statement, an explanation of each viewpoint, and your conclusion. Independent research of Christian and secular viewpoints will be required to complete this activity.

Answers in Genesis (www.answersingenesis.org), the Creation Research Society (www.creationresearch.org), and the Institute for Creation Research (www.icr.org) are recommended resources.

Name

Word to Know – Write the definition in the space provided below.

Volcanism

Escarpment

Questions

NOTE: This week's questions are the same as the questions in the text. It is important that the main concept of each question is expressed in the answer. The answers should be worded in a way that shows the student understands and can apply them to the answer rather than simple memorization.

1. How was Grand Canyon formed?

2. What causes the geysers in Yellowstone Park?

3. How did Niagara Falls form?

4. Why are the Appalachian and Rocky Mountains so different?

5. How long does it take to form petrified wood?

6. How are stalactites and stalagmites formed?

7. How is coal formed?

8. How is natural gas formed?

9. How is oil formed?

10. Are dinosaur fossils the most abundant type of fossils?

Quizzes & Tests Section

Q	Archaeology Book / Concepts & Comprehension	Quiz 1	Scope: Chapters 1-2	Total score: ____of 100	

Name

Define: (5 Points Each Answer)

1. accession year: _____

2. AD: _____

3. BC: _____

4. carbon dating: _____

5. EB: _____

6. LB: _____

7. MB: _____

8. baulk: _____

9. synchronism: _____

10. mastabas: _____

Multiple Answer Questions: (2 Points Each Blank)

11. What are the four main periods of archaeological time?

    a. _____   c. _____

    b. _____   d. _____

12. For what three reasons were cities built on hills?

    a. _____

    b. _____

    c. _____

Short Answer Questions: (4 Points Each Question)

13. What does the word *archaeology* mean?_____

_____

14. When did people first start using coins? _____

_____

15. What was the name of the Egyptian god of the Nile River? _____

_____

16. What is the Egyptian name for Egypt? _____

_____

17. Who was the first Egyptian king to build a pyramid? _____

18. Who built the biggest pyramid in Egypt? _____

_____

Applied Learning Activity: (12 Points Total; 1 Point Each Answer)

19. Identify the Pyramids, Temples, Tombs, and unique features on Giza Map:

Pyramid of Kufu

Valley Temple of Kufu

Pyramid of Menkaure

Valley Temple of Menkaure

Pyramid of Kahfre

Valley Temple of Kahfre

The Sphinx

The Temple of the Sphinx

Pyramid of Queens

Queen's Tombs

Eastern Cemetery

Mortuary Temple

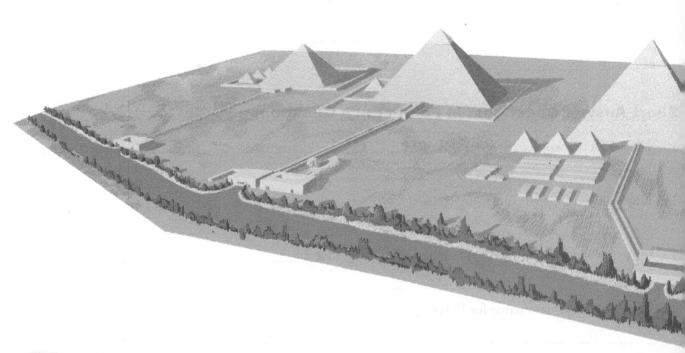

Name

Define: (5 Points Each Answer)

1. amphitheater: _____

2. Anatolia: _____

3. centurion:_____

4. Chaldees: _____

5. bulla: _____

6. scarab: _____

Multiple Answer Questions: (4 Points Each Blank)

7. What two nations did the Syrians think had come to attack them?

   a. _____   b. _____

Short Answer Questions: (4 Points Each Question)

8. Which was the strongest nation in the Middle East 3000 years ago?_____

   _____

9. Who were the Hittites descended from? _____

   _____

10. How often were the Hittites mentioned in the King James Version of the Bible? _____

    _____

11. Who wrote the book *The Empire of the Hittites*? _____

    _____

12. In the Bible, how many references are there to Ur of the Chaldees? _____

13. Why did Woolley not excavate the cemetery as soon as he found it?_____

    _____

14. What was the name of the people who occupied ancient Ur? _____

15. Who discovered Nineveh? _____

16. What was the name of the ruins where Layard first started digging? _____

17. What was the name of the king of Israel that was mentioned on the black pillar Layard found in Nimrud?_____

18. What was the name of the king of Israel when Sennacherib besieged Jerusalem?_____

Applied Learning Activity: (2 Point Each Blank)

19-21. Identify the writing materials and answer the questions: **Vellum, Papyrus, Pottery**

a. _____   b. _____   c. _____

22. What was Vellum made from? _____
_____

23. What is the name of a person who made Vellum? _____

24. What was Papyrus made from? _____
_____

25. Who made Papyrus and sold it all over the Mediterranean? _____

26. What was the main city for Papyrus production? _____

27. What word to we get from this city? _____

Q	Archaeology Book	Quiz 3	Scope:	Total score:	
	Concepts & Comprehension		Chapters 6-8	____ of 100	

Name

Define: (5 Points Each Answer)

1. cuneiform: _____

2. strata: _____

3. syncline: _____

4. Persia: _____

5. rhyton: _____

6. cistern: _____

7. Nabataeans: _____

8. wadi: _____

Short Answer Questions: (4 Points Each Question)

9. What was the name of the cuneiform record which told a story similar to the Bible record of Noah and the flood? _____
_____

10. Which Assyrian king compiled a library of tablets in Nineveh?_____

11. Which king made Babylon a city of gold?_____

12. Which Bible prophet predicted that Babylon would become uninhabited?_____

13. Who was the king who first carved out the Medo-Persian Empire?_____

14. In what year did he conquer Babylon? _____

15. Which Persian king left an inscription on the rock face of the Zagros Mountain near Bisitun?_____

16. What was the name of the great Persian city that Darius built?_____

17. Which Bible prophet wrote a book about Petra?_____

18. Which Roman emperor had a road constructed through Petra? _____

Applied Learning Activity: (20 Points)

In your own words, tell the story of Esther. Include by name at least four of the characters and the name of the Jewish feast still celebrated today to commemorate the deliverance. (You may use the back of this page if more room is needed.)

Name

Define: (5 Points Each Answer)

1. Baal: _____

2. causeway: _____

3. Yehovah: _____

4. scroll: _____

5. annunciation: _____

6. Calvary: _____

7. Golgotha: _____

8. grotto: _____

9. Messiah: _____

10. ossuary: _____

Multiple Answer Questions: (1 Point Each Blank)

11. What were the four main cities of ancient Phoenicia?

    a. _____   c. _____

    b. _____   d. _____

Short Answer Questions: (4 Points Each Question)

12. Which trees were Phoenicia famous for? _____

13. Whose tomb did Pierre Montet find? _____

14. Which Bible prophet challenged the prophets of Baal? _____

15. Which Bible prophet predicted that ancient Tyre would never be found? _____

16. In what year was the first Dead Sea Scroll found? _____

17. What was the name of the settlement near the cave where the Dead Sea Scrolls were found? _____

_____

18. Which Roman Emperor adopted Christianity as the state religion? _____

19. Identify the languages: **Sumerian, Phoenician, Hebrew, Egyptian**

_____

_____

_____

_____

20. Name two books of the Bible that include chapters written in acrostic form (a form of Hebrew poetry):

   a. _____

   b. _____

Name

Define: (3 Points Each Answer)

1. carbon dating: _____

2. baulk: _____

3. synchronism: _____

4. mastabas: _____

5. centurion: _____

6. Chaldees: _____

7. bulla: _____

8. cuneiform: _____

9. syncline: _____

10. Persia: _____

11. rhyton: _____

12. annunciation: _____

13. ossuary: _____

14. grotto: _____

Multiple Answer Questions: (1 Point Each Blank)

15. What are the four main periods of archaeological time?

   a. _____   c. _____

   b. _____   d. _____

16. What are two reasons cities were built on hills?

   a. _____   b. _____

17. What were the four main cities of ancient Phoenicia?

   a. _____   c. _____

   b. _____   d. _____

Short Answer Questions: (3 Points Each Question)

18. Who was the first Egyptian king to build a pyramid? _____

19. Who built the biggest pyramid in Egypt? _____

20. What was the name of the king of Israel that was mentioned on the black pillar Layard found in Nimrud? _____

21. What was the name of the king of Israel when Sennacherib besieged Jerusalem? _____
_____

22. Which Bible prophet predicted that Babylon would become uninhabited? _____

23. Who was the king who first carved out the Medo-Persian Empire? _____

24. Which Bible prophet predicted that ancient Tyre would never be found? _____

25. In what year was the first Dead Sea Scroll found? _____

Applied Learning Activity: (1 Point Each Answer)

26. Identify the Pyramids, Temples, Tombs, and unique features on Giza Map:

Pyramid of Kufu

Valley Temple of Kufu

Pyramid of Menkaure

Valley Temple of Menkaure

Pyramid of Kahfre

Valley Temple of Kahfre

The Sphinx

The Temple of the Sphinx

Pyramid of Queens

Queen's Tombs

Eastern Cemetery

Mortuary Temple

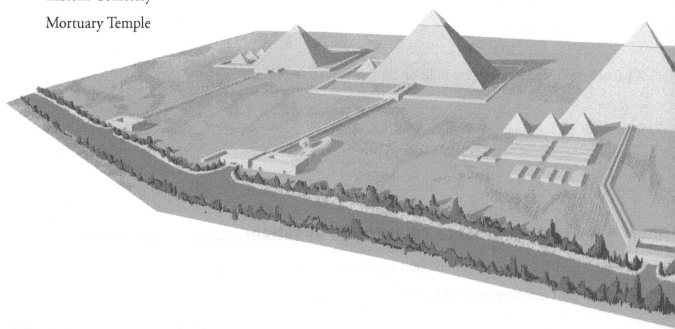

27. Identify the languages: **Sumerian, Phoenician, Hebrew, Egyptian**

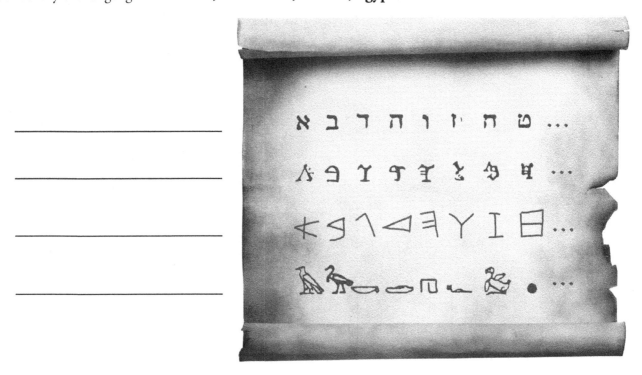

_____

_____

_____

_____

28. Name two books of the Bible that include chapters written in acrostic form (a form of Hebrew poetry): (2 Points Each Answer)

a. _____

b. _____

Q	Geology Book	Quiz 1	Scope: Chapters 1-3	Total score: ____of 100	
	Concepts & Comprehension				

Name

Define: (5 Points Each Answer)

1. Principle of uniformity : _____

2. Principle of catastrophe: _____

3. Sediment: _____

4. Metamorphism: _____

Multiple Answer Questions: (2 Points Each Blank)

5. There are two ways of thinking about the unobserved past. What are they? (3 Points Each Answer)

    a. _____

    b. _____

6. In what order did God create the heavens and earth?

    Day 1. _____   Day 4. _____

    Day 2. _____   Day 5. _____

    Day 3. _____   Day 6. _____

7. What are the main "zones" into which the earth is divided?

    a. _____   d. _____

    b. _____   e. _____

8. List the 3 types of plateaus and give an example of each.

    a. _____   b. _____

    c. _____   d. _____

    e. _____   f. _____

9. List the 4 types of mountains and give an example of each type.

    a. _____   b. _____

    c. _____   d. _____

    e. _____   f. _____

    g. _____   h. _____

10. List which category the following types of rocks belong to.

Granite   a._____

Marble   b._____

Shale   c._____

Limestone d._____

Coal   e._____

Rhyolite   f._____

Slate   g._____

Name _____

Define: (5 Points Each Answer)

1. Erosion: _____

2. Petrification: _____

3. turbidite: _____

4. Gastrolith: _____

5. Fumaroles: _____

6. Carbon dating: _____

Multiple Answer Questions: (2 Points Each Blank)

7. List the 5 primary causes of normal erosion.

   a. _____

   b. _____

   c. _____

   d. _____

   e. _____

8. Name four different types of fossils.

   a. _____   c. _____

   b. _____   d. _____

9. What are unstable atoms called? What is the most well-known radioactive atom?

   a. _____

   b. _____

Short Answers: (5 Points Each Question)

10. Sand is made primarily of what mineral? _____

_____

11. How is a turbidite formed? _____

_____

12. What is the main condition required for a fossil to form? _____

_____

13. What factors influence whether a rock will bend or break? _____

_____

Applied Learning Activity: (7 Points Each Question)

14. Describe how sedimentary rocks are formed.

15. Describe how metamorphic rocks are thought to form.

16. Explain the process of carbon dating in determining when a plant died.

17. In your own words, write a description of how dinosaur fossils were formed.

Q	Geology Book — Concepts & Comprehension	Quiz 3	Scope: Chapter 5-6	Total score: ____of 100

Name _____

Define: (5 Points Each Answer)

1. Magnetic field: _____

2. Uplift: _____

3. Second law of science: _____

4. Fountains of the deep: _____

5. Glacier: _____

6. Polar ice cap: _____

Multiple Answer Questions: (2 Points Each Blank)

7. What four events have had the greatest impact in shaping the earth's geology?

    a. _____

    b. _____

    c. _____

    d. _____

8. What was the cause of the global Flood? What were the geological results of the Flood?

    a. _____

    b. _____

9. List the 5 ways to date the earth discussed in the book.

    a. _____

    b. _____

    c. _____

    d. _____

    e. _____

Short Answers: (6 Points Each Question)

10. What role has the Fall played in shaping today's earth? _____

_____

11. What conclusion can be drawn about the age of the earth from the various dating methods? _____

_____

12. How did the creation event affect the earth's geology? _____

_____

13. Why do we find ocean fossils near the top of Mt. Everest? _____

_____

14. What caused the Ice Age? _____

_____

Applied Learning Activity: (18 Points)

15. Based on your previous essay, explain the difference between the secular view and the Biblical view of the cause of the Ice Age.

Q	Geology Book	Quiz 4	Scope:	Total score:
	Concepts & Comprehension		Chapter 7-8	____of 100

Name

Define: (5 Points Each Answer)

1. Volcanism: _____

2. Escarpment: _____

Short Answers: (5 Points Each Question)

3. How was Grand Canyon formed?_____

_____

4. What causes the geysers in Yellowstone Park? _____

_____

5. How did Niagara Falls form? _____

_____

6. Why are the Appalachian and Rocky Mountains so different? _____

_____

7. How long does it take to form petrified wood?_____

_____

8. How are stalactites and stalagmites formed? _____

_____

9. How is coal formed?_____

_____

10. How is natural gas formed? _____

_____

11. How is oil formed?_____

_____

12. Are dinosaur fossils the most abundant type of fossils? _____

_____

Applied Learning Activity: (20 Points Each Question)

13. As part of God's judgment for disobedience at the end of time, the earth will undergo heat waves, droughts, flaming comets, earthquakes and plagues. Read Psalm 46 and explain in your own words how this passage of scripture explains the hope that Christians have in times of hardship on the earth.

14. Read 2 Peter 3:10-13 and Revelation 21:1-4. Based on these verses, explain in your own words the hope that Christians have for eternity. Where will you be throughout eternity?

T	Geology Book
	Concepts & Comprehension

Test	Scope: Chapters 1-8	Total score: ____ of 100

Name

Define: (2 Points Each Answer)

1. Principle of uniformity: _____

2. Principle of catastrophe: _____

3. Erosion: _____

4. Petrification: _____

5. turbidite: _____

6. Gastrolith: _____

7. Fumaroles: _____

8. Metamorphism: _____

9. Magnetic field: _____

10. Sediment: _____

11. Second law of science: _____

12. Fountains of the deep: _____

13. Glacier: _____

14. Volcanism: _____

Multiple Answer Questions: (2 Points Each Blank)

15. There are two ways of thinking about the unobserved past. What are they?

   a. _____

   b. _____

16. What are the main "zones" into which the earth is divided?

   a. _____

   b. _____

   c. _____

   d. _____

17. What four events have had the greatest impact in shaping the earth's geology?

   a. _____

   b. _____

   c. _____

   d. _____

Short Answer Questions: (2 Points Each Question)

18. What is the main condition required for a fossil to form? _____
_____

19. What factors influence whether a rock will bend or break? _____
_____

20. What conclusion can be drawn about the age of the earth from the various dating methods? _____
_____

21. How did the creation event affect the earth's geology?_____
_____

22 Why do we find ocean fossils near the top of Mt. Everest? _____
_____

23. What caused the Ice Age? _____
_____

24. What was the cause of the global flood? _____
_____

25. How was Grand Canyon formed?_____
_____

26. How did Niagara Falls form? _____
_____

27. Why are the Appalachian and Rocky Mountains so different? _____
_____

28. How long does it take to form petrified wood? _____
_____

Applied Learning Activity: (3 Points Each Answer)

29. List which category the following types of rocks belong to.

Granite   a._____

Marble   b._____

Shale   c._____

Limestone d._____

Coal   e._____

Rhyolite   f._____

Slate   g._____

30. Describe how sedimentary rocks are formed.

31. Describe how metamorphic rocks are thought to form.

32. Describe the process of carbon dating in determining when a plant died.

Answer Keys

Archaeology Book - Worksheet Answer Keys

Chapter 1 – What Archaeology is All About – Worksheet 1

accession year — the year a king actually began his reign

AD — Anno Domini (the year of our lord); the years after the Christian era began

Archaeology — study of beginnings

Artifact — an item from antiquity found in an excavation

BC — Before Christ; the years before the Christian era began

carbon dating — calculating the amount of carbon left in organic material that has died

ceramic — something made of pottery

chronology — time periods; dates in which events happened

debris — discarded rubbish

EB — the Early Bronze Period

exile — a people sent out of their home country to another country

exodus — going out; applied to the Israelites leaving Egypt

hieroglyphs — Egyptian picture writing

LB — the Late Bronze Period

MB — the Middle Bronze Period

millennium — one thousand years

non-accession year — the first complete year of a king's reign

pottery — a vessel made of clay fired in a kiln

synchronism — something happening at the same time

tell — a Hebrew word meaning "ruins" applied to hills on which people once lived

1. A study about beginnings

2. Defense, heat, and floods

3. 600 BC

4. It helps them identify from which period the pottery comes.

5. Early Bronze, Middle Bronze, Late Bronze, Iron Age.

Chapter 2 – Land of Egypt – Worksheet 1

Asiatic — in Egyptian terms, someone from Syria or Palestine

baulk — the vertical ridge left between two excavated squares in the ground

dowry — gift given to a prospective bride at the time of her marriage

drachma — a Greek coin worth about a day's wages

dynasty — a succession of kings descended from one another

mastabas — mud-brick structures beneath which were tomb chambers

Nubia — a country south of Egypt now called Sudan

Pharoah — title applied to many Egyptian kings

1. Misr

2. Zoser

3. Khufu

4. Hapi

5. Straw

Chapter 3 – The Hittites – Worksheet 1

amphitheater — a circle of seats surrounding an area where gladiators fought each other or fought wild beasts

Anatolia — mountainous area in central Turkey

bathhouse — a club where citizens could bathe in cold, warm, or hot water

inscription — writing made on clay, stone, papyrus, or animal skins

1. The Hittites

2. Hittites and Egyptians

3. Heth

4. Forty-six

5. William Wright

Chapter 4 – Ur of the Chaldees – Worksheet 1

centurion — a military officer in charge of a hundred men

Chaldees — people who used to live in southern Iraq

nomad — a person who lived in a tent that could be moved from place to place

papyrus — sheets of writing material made from the Egyptian papyrus plant

1. Four

2. Sir Leonard Woolley

3. He wanted to learn more about Ur before he excavated such an important site.

4. Sumerians

5. Evidence of human sacrifice

Chapter 5 – Assyria – Worksheet 1

bulla — an impression made on clay with a seal (plural: bullae)

Medes — people who used to live in northern Iran

scarab — model of a dung beetle with an inscription engraved on it for sealing documents

seal — an object made of stone, metal, or clay with a name engraved on it used to impress in soft clay

1. Henry Austin Layard

2. Nimrud

3. Jehu

4. Hezekiah

5. Forty-six

Chapter 6 – Babylon: City of Gold – Worksheet 1

Armenians — people who lived in eastern Turkey and northern Iraq

cuneiform — a form of writing using a wedge-shaped stylus to make an impression on a clay tablet

strata — a layer of occupation exposed by excavations

syncline — a boat-shaped geological formation

1. The Gilgamesh Epic

2. Ashur-Bani-Pal

3. Asphalt

4. Nebuchadnezzar

5. Isaiah

Chapter 7 – Persia – Worksheet 1

Persia — a country in central Iran

rhyton — a drinking vessel shaped like a human or animal

1. Cyrus the Great

2. 539 BC

3. Darius the Great

4. Persepolis

5. Haman

Chapter 8 – Petra – Worksheet 1

bedouin — Arabs living in tents with no fixed address

cistern — a hole dug in rock to store rainwater

Edom — country in southern Jordan

Edomites — people descended from Edom, also known as Esau, Jacob's brother

Nabataeans — people descended from Nabaioth, who occupied Petra

siq — narrow valley between two high rock formations

theater — a stage for actors in front of which was a semi-circle of seats

wadi — a dry riverbed, carrying water only when it rained

1. 1812 AD

2. Esau's

3. Edomites

4. Obadiah

5. Trajan

Chapter 9 – The Phoenicians – Worksheet 1

Baal — a word meaning "lord" and the name of a Phoenician god

causeway — a built-up road

Yehovah — a Hebrew name for God, usually spelled Jehovah, but there is no "J" in the Hebrew alphabet

1. Gebal, Berytus Sidon and Tyre
2. Cyprus
3. Ahiram
4. Elijah
5. Ezekiel

Chapter 10 – The Dead Sea Scrolls – Worksheet 1

Scroll — papyrus or animal skin document rolled up into a cylinder

vellum — animal skin treated to be used as writing material

1. 1947
2. 22
3. Vellum
4. Psalm 119
5. Qumran

Chapter 11 – Israel – Worksheet 1

annunciation — an announcement

Calvary — Latin word meaning "skull"

Golgotha — Hebrew word meaning "skull"

grotto — cave

Messiah — meaning "Anointed One" and applied to an expected Jewish leader

ossuary — a box in which human bones were preserved

Passover — Jewish ceremony celebrating the Exodus from Egypt

1. Constantine
2. Yehovah saves
3. Jordan River
4. Capernaum
5. Skull

Geology Book - Worksheet Answer Keys

Introduction & Chapter 1 – Planet Earth – Worksheet 1

Principle of uniformity — the scientific thought that past proccesses are no different than processes today, meaning everything happens by gradual process over very long periods of time

Principle of catastrophe — the scientific thought that sees evidence of rapid, highly energetic events over short periods of time, doing a lot of geologic work

Asthenosphere — a suspected area in the uppermost portion of the earth's mantle where material is hot and deforms easily

Plate — huge regions of the earth identified by zones of earthquake activity

1. Origins science cannot be studied with repeatable, observable experiments in the present.

2. Uniformity (the present is the key to the past) and catastrophe (highly energetic events operated over short periods of time and did much geologic work rapidly)

3. In the Bible

4. Day 1: Earth, space, time, light; Day 2: Atmosphere; Day 3: dry land, plants; Day 4: sun, moon, stars, planets; Day 5: sea and flying creatures; Day 6: land animals, people

5. Sin can be defined as rebellion against God

6. Crust, mantle, outer core, inner core

7. Continental crust (composed of granitic rock covered by sedimentary rock); oceanic crust (composed primarily of basaltic rock)

8. Provides the air we breathe, protects us from harmful cosmic radiation, and gives us weather.

Chapter 2 – The Ground We Stand Upon – Worksheet 1

Igneous rocks — rock formed when hot, molten magma cools and solidifies

Sedimentary rock — rock formed by the deposition and consolidation of loose particles of sediment, and those formed by precipitation from water

Metamorphic rock — rocks formed when heat, pressure and/or chemical action alters previously existing rock

Ripple Marks — marks which indicate moving water flowed over a rock layer when the sediments were still muddy and yet to harden

Crossbed — areas of extremely large ripple marks

Concretions — concreted masses of sedimentary rock that has been eroded out of a softer area of rock

Metamorphism — a process of heat and pressure that causes one rock to alter into another

Type	Catagory	Composition	Formation	Found
Granite	Igneous	Quartz and feldspar with mica and hornblende	Formed when molten rock is cooled	Mountains Upper mantle
Rhyolite	Igneous	Quartz and feldspar with mica and hornblende	Formed when molten rock erupts on land and solidifies	Land

Type	Catagory	Composition	Formation	Found
Obsidian	Igneous	Quartz and feldspar with mica and hornblende	Formed by the rapid cooling of lava as it flows in the surface of the ground	Land
Pumice	Igneous	Quartz and feldspar with mica and hornblende	Formed by eruptions on land—the cooling process forms air pockets in the rock	Land
Basalt	Igneous	Pyroxene, plagioclase feldspar	Solidified molten lava under water and on land	Oceanic crust, land
Shale	Clastic Sedimentary	Cemented particles of clay (and minor silt)	Formed from previously existing rocks which were eroded, transported and redeposited elsewhere	Mountains, land
Sandstone	Clastic Sedimentary	Quartz sand, particles big enough to be seen	Formed from previously existing rocks which were eroded, transported and redeposited elsewhere	Mountains, land
Conglomerate	Clastic Sedimentary	Pebble-size to boulder-size grains mixed with smaller sand or clay particles	Formed from previously existing rocks which were eroded, transported and redeposited elsewhere	Mountains, land
Limestone	Organic chemical sedimentary	Calcium carbonate from shells of sea creatures, reef fragments or limey secretions of sea creatures	Formed when water can no longer keep various chemicals dissolved within it	Sea floors, land
Diatomaceous earth	Organic chemical sedimentary	Collection of shells from diatoms or radiolarians and certain algae	Formed when water can no longer keep various chemicals dissolved within it	Land
Coal	Organic chemical sedimentary	Buried plant material	Formed when water can no longer keep various chemicals dissolved within it	Land

Type	Catagory	Composition	Formation	Found
Limestone	Inorganic chemical sedimentary	Calcium carbonate derived from inorganic sources	Formed when water can no longer keep various chemicals dissolved within it	Caves, mineral springs, stalactites, stalagmites
Dolomite	Inorganic chemical sedimentary	Calcium carbonate with magnesium atoms	Formed when water can no longer keep various chemicals dissolved within it	Land
Evaporites	Inorganic chemical sedimentary	The remains of evaporated seawater	Some were formed when a huge volume of mineral-laden water came up through the ocean floor basalts and released its dissolved content when it hit the cold ocean waters	Land
Slate	Metamorphic	Shale	Shale subjected to heat and pressure	Land
Schist	Metamorphic	Shale	Slate that continues to undergo heat and pressure	Land
Gneiss	Metamorphic	Alternating bands of different minerals from other sedimentary or igneous rocks	Formed from other sedimentary or igneous rocks that have been subjected to heat and pressure	Land
Quartzite	Metamorphic	Quartz sandstone	Quartz sandstone that has been subjected to change	Land
Marble	Metamorphic	Limestone	Heat and pressure applied to limestone	Land

Chapter 3 – The Earth's Surface – Worksheet 1

Plain — a broad area of relatively flat land

Sediment — a natural material broken down by processes of erosion and weathering; can be transported or deposited by water or wind

Plateau — flat lying sediment layers similar to plains but at higher elevations

Mountain — a large landform rising abruptly from the surrounding area

Canyon — a deep ravine between cliffs often carved by streams or rivers

1. Because the sediment deposited there is rich in nutrients

2. Fault: rock is broken and shoved up (Colorado Plateau); warped: regional squeezing or slow uplift (Appalachian mountains); lava: hardened lava plains that may have been uplifted or hardened at the current level (Columbia River basalts)

3. Folded: layers of sediments that have been crumpled by pressures from the side (Alps, Himalayas, Appalachians, Rocky Mountains); Domed: sediments pushed up from below (Black Hills of South Dakota); fault block: one area of sediments are pushed up (Grand Teton Mountains); volcanic: molten lavas pushed out to the surface of the earth (Hawaii's volcanic islands, Mount Rainier, Mount St. Helens, Mount Ararat)

4. Erosion

Chapter 4 – Erosion/Deposition – Worksheet 1

Erosion — the process by which soil and rock are worn away

Deposition — the process by which sediments are deposited onto a landform

Turbidite — an underwater rapid deposition of mud which hardens into a layer of rock formed by mud

1. Rain, ice, plants and animals, chemicals, ocean waves

2. Cavitation occurs when tiny bubbles in moving water explode inwardly; plucking is where rocks are picked up by moving water; kolk is like an underwater tornado that breaks up rock.

3. First, an event such as an earthquake starts a mud flow underwater. Next, the mud flow spreads out. Eventually the mud flow hardens into a layer of rock.

4. Flooding and tidal waves or tsunamis

Chapter 4 – Sediments/Fossilization – Worksheet 2

Compaction — a process in the formation of sedimentary rock when the materials are pushed together tightly, leaving little to no open spaces

Cementation — a process in the formation of sedimentary rock when minerals are dissolved which then help to solidify the rock by acting as glue or cement

Fossils — the remains of plants and animals that were once alive

Petrification — the process by which trees, plants, and even animals are solidified by burial in hot, silica-rich water

Gastrolith — rounded stones which were used by plant-eating dinosaurs to aid in digestion and sometimes found with fossilized remains

Coprolite — fossilized animal or dinosaur dung; can be used to determine a creature's diet

1. Silica

2. First, layers of sediment are deposited. The weight of the water and the sediments on top begin to compact the sediments underneath. Next, warm water circulates throughout the sediments and dissolves certain minerals. The dissolved minerals surround the individual grains of sediment. Finally, when the water cools off and stops moving, the dissolved minerals act as a "glue" that cements the grains of

sediment together to form sedimentary rock.

3. The organism must be buried rapidly, protected from scavengers and from decomposition by bacteria and chemicals.

4. Hard parts are preserved; replacement by other minerals; cast or mold are all that remains; petrification; cabonization; preservation of soft parts; frozen animals; animal tracks and worm burrows; coprolites; gastroliths

5. First, if a dinosaur was not on the ark, then it drowned in the Great Flood. Next, the animal was buried rapidly as the flood deposited soft layers of material that later hardened into stone. Then, a process of fossilization occurred, such as the bones being replaced by dissolved minerals in the ground water. Finally, the fossils became exposed as the ground around the animal eroded away.

Chapter 4 – Volcanism/Deformation of Rocks/Continents – Worksheet 3

Volcanism — the eruption of molten rock (magma) onto the surface of the earth

Fumaroles — an opening in the earth's crust, usually associated with volcanic activity

Geyser — underground water that has been heated to an excessive degree and because of pressure bursts out of the ground temporarily

Fault — A fracture in rock along which separation or movement has taken place.

Earthquake — a sudden release of energy below the earth's crust which cause the earth's crust to move or shake.

1. Some volcanoes erupt by just spilling lava out from their top; others explode out of their top.

2. In a normal fault, the hanging wall moves downward relative to the foot wall. In a reverse fault, the hanging wall moves upward. In a strike-slip fault, both walls move sideways.

3. Whether it is soft or brittle, how deep it is buried

4. Answers will vary.

Chapter 4 – Metamorphism/Radioisotope Decay – Worksheet 4

Atom — the basic component of chemical elements

Isotope — variations of an element's atoms, usually in the different number of neutrons

Radioisotope dating — the process of using the rate of atomic decay to determine how old an object may be

Carbon dating — A process which uses the decay of carbon 14 to estimate the age of things that were once living

1. Heat and pressure recrystallize the minerals in rock into new mineral combinations. Some believe it happened over long periods of time; others believe it happened over short periods of time.

2. Radioactive; uranium

3. Uranium is used by nuclear power plants to generate electricity

4. The daughter isotope is formed from the decay of the parent isotope.

5. When a plant is living, it takes the isotope carbon 14 into its leaves, stems, and seeds. After the plant

dies, the carbon 14 decays into nitrogen 14. Scientists can measure the amounts of both carbon 14 and carbon 12. Since they know the time it takes the isotope to decay, they can calculate when the plant died.

Chapter 5 – Ways to Date the Entire Earth – Worksheet 1

Magnetic field — A field that exerts forces on objects made of magnetic materials; made up of many lines of force

Uplift — in geology, a tectonic uplift is a geological process most often caused by plate tectonics which causes an increase in elevation

1. Measuring the chemicals in the ocean, measuring the rate of erosion of the continents, measuring the sediments in the ocean, dating the atmosphere, dating the magnetic field

2. We must consider the possibility of processes happening at different rates. We can measure the rate that certain processes currently happen, but a massive flood or other event could have had a major impact in a very short amount of time

3. A majority of methods used to age-date the earth yield ages far less than the acclaimed billions of years.

Chapter 6 – Great Geologic Events of the Past – Worksheet 1

Second law of science — also referred to as the second law of thermodynamics, which states that in every process or reaction in the universe the components deteriorate

Fountains of the deep — a phrase mentioned in Genesis 7 as a reference to sources of water as part of the Great Flood of Noah.

Glacier — a huge mass of ice that moves slowly over land

Polar ice cap — a high latitude region of a planet that is covered in ice

1. Creation, the Fall, Flood, Ice Age

2. Formed the cores of the continents, some erosion and deposition probably happened

3. The Bible says in Genesis 3 that the entire creation came under the curse of sin, including plants, animals, mankind, and the earth. As a result of the curse, everything is wearing down and deteriorating.

4. God sent the Flood as a judgment on the wickedness of mankind. A worldwide flood would have caused a vast change to the earth's surface. The continents were uplifting, mountain ranges and lakes were formed and rock and fossil layers were laid down.

5. The top of Mount Everest was once underwater and was later pushed up after the Flood waters receded.

6. See page 67 of *The Geology Book*

Chapter 7 – Questions People Ask – Worksheet 1

Volcanism — this is the process by which molten rock or lava erupts through the surface of a planet

Escarpment — a steep slope or long cliff that occurs from erosion or faulting and separates two relatively level areas of differing elevations

1. Many geologists now recognize that the Colorado River cound not have carved the Grand Canyon. At the end of Noah's flood it appears that a great volume of water was trapped, held in place by the Kaibab

Upwarp. Ice Age rains filled the lake to overflowing and as it burst through its mountain "dam," the huge volume of lake waters carved the canyon.

2. At Yellowstone Park, the soil and rock is thin, allowing very hot material to be near the surface. As rain and run-off water trickle down into the earth they get heated. In some places the underground water is trapped and when heated to an excessive degree, it bursts out in a geyser. The geyser stops once the pressure is relieved, but will erupt again as building pressures exceed the maximum.

3. The level of water in Lake Erie is somewhat higher than the elevation of nearby Lake Ontario. A river draining the waters of Lake Erie into Lake Ontario runs over the Niagara escarpment, resulting in a spectacular set of falls. Erosion takes place as the water roars over the falls, and the escarpment naturally recedes toward Lake Erie.

4. Both mountain chains are the result of layers of sediments deposited by Noah's flood. The Appalachians buckled up in the early stages of the flood and were subjected to massive erosion by the continuing flood waters. The Rocky Mountains buckled up late in the flood, extending up above the flood waters as the waters drained off. Thus, the erosion to which they were subjected was much less intense.

5. Petrified wood can form, under laboratory conditions, in a very short period of time. The speed of petrifaction is related to the pressures which inject the hot silica-rich waters into the wood.

6. When water saturated with calcium carbonate enters an open space such as a cave, it cools off or evaporates, leaving the calcium carbonate behind. Stalactites (holding "tightly" to the ceiling) and stalagmites (which are usually larger and thus more "mighty" than stalactites) are formed as this calcium carbonate precipitates out of the water.

7. Coal can be formed in a laboratory by heating organic material away from oxygen but in the presence of volcanic clay. Under such conditions, coal can be formed in a matter of hours. One wonders if the abundant forest growing before the flood would not have formed huge log mats floating on the flood ocean. As these decayed and were buried by hot sediments in the presence of volcanic clay, they might have rapidly turned to coal.

8. Natural gas, mostly methane, is given off in the coalification process. The largest quantities of it, however, are found in deep rocks not associated with decomposition of organic material. Evidently, some natural gas is from both organic and inorganic sources.

9. Many theories have been promoted as to the specific origin of oil. The best seems to be that it is the remains of algae once floating in the ocean but buried in ocean sediments. Oil is not the remains of dinosaurs as has sometimes been claimed.

10. No - they are rare. Most fossils are of sea creatures, fish, and insects. Only relatively few fossils are of land animals, specifically dinosaurs.

Archaeology - Unit Quiz Answer Key

Unit One Quiz, chapters 1-3

1. **accession year** — the year a king actually began his reign
2. **AD** — Anno Domini (the year of our lord); the years after the Christian era began
3. **BC** — Before Christ; the years before the Christian era began
4. **carbon dating** — calculating the amount of carbon left in organic material that has died
5. **EB** — the Early Bronze Period
6. **LB** — the Late Bronze Period
7. **MB** — the Middle Bronze Period
8. **baulk** — the vertical ridge left between two excavated squares in the ground
9. **synchronism** — something happening at the same time
10. **mastabas** — mud-brick structures beneath which were tomb chambers
11. Early Bronze, Middle Bronze, Late Bronze, Iron Age
12. Defense, heat, and floods
13. A study about beginnings
14. 600 BC
15. Hapi
16. Misr
17. Zoser
18. Khufu
19. Identify the Pyramids, Temples, Tombs, and unique features on Giza Map:

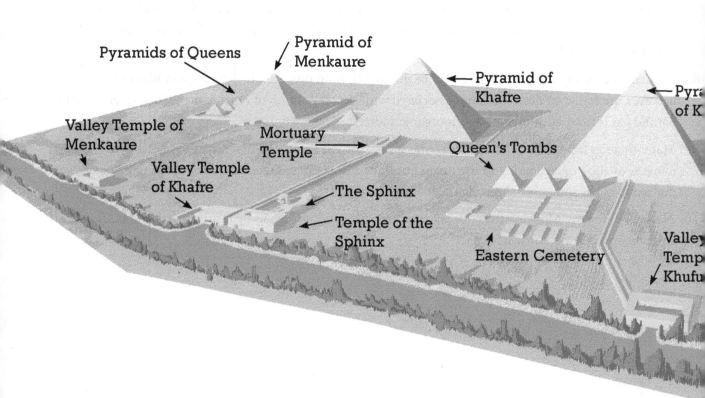

1. **amphitheater** — a circle of seats surrounding an area where gladiators fought each other or fought wild beasts

2. **Anatolia** — mountainous area in central Turkey

3. **centurion** — a military officer in charge of a hundred men

4. **Chaldees** — people who used to live in southern Iraq

5. **bulla** — an impression made on clay with a seal (plural: bullae)

6. **scarab** — model of a dung beetle with an inscription engraved on it for sealing documents

7. The Syrians thought the Hittites and Egyptians had come to attack them.

8. The strongest nation in the Middle East 3000 years ago was the Hittites.

9. The Hittites were descended from Heth.

10. The Hittites were mentioned forty-six times in the King James Version of the Bible.

11. William Wright wrote the book "The Empire of the Hittites."

12. In the Bible, there are four references to "Ur of the Chaldees."

13. He wanted to learn more about Ur before he excavated such an important site.

14. Sumerians was the name of the people who occupied ancient Ur.

15. Henry Austin Layard discovered Nineveh.

16. Nimrud was the name of the ruins where Layard first started digging.

17. Jehu was the name of the king of Israel that was mentioned on the black pillar Layard found in Nimrud.

18. Hezekiah was the name of the king of Israel when Sennacherib besieged Jerusalem.

19-21. Identify the writing materials and answer the questions: **Vellum, Papyrus, Pottery**

a. Pottery   b. Vellum   c. Papyrus

22. Vellum made from leather (animal skins that were scraped clean and treated for preservation)

23. A person who made Vellum was called a tanner.

24. Papyrus was made from papyrus stalks from Egypt.

25. The Phoenicians, now Lebanese, made Papyrus and sold it all over the Mediterranean.

26. Byblos was the main city for Papyrus production.

27. We get the word "Bible" from this city.

Unit Three Quiz, chapters 6-8

1. **cuneiform** — a form of writing using a wedge-shaped stylus to make an impression on a clay tablet

2. **strata** — a layer of occupation exposed by excavations

3. **syncline** — a boat-shaped geological formation

4. **Persia** — a country in central Iran

5. **rhyton** — a drinking vessel shaped like a human or animal

6. **cistern** — a hole dug in rock to store rainwater

7. **Nabataeans** — people descended from Nabaioth who occupied Petra

8. **wadi** — a dry riverbed, carrying water only when it rained

9. The Gilgamesh Epic

10. Ashur-Bani-Pal

11. Nebuchadnezzar

12. Isaiah

13. Cyrus the Great

14. 539 BC

15. Darius the Great

16. Persepolis

17. Obadiah

18. Trajan

19. **Applied Learning Activity:** (20 Points-4 Points for each character and for Purim)

Student should include by name at least four of the characters: **Darius, Xerxes, Vashti, Esther, Haman, Mordecai** and describe their role in the account. The story should reflect the Biblical account of Esther.

Purim is the name of the Jewish feast still celebrated today to commemorate the deliverance.

Unit Four Quiz, chapters 9-12

1. **Baal** — a word meaning "lord" and the name of a Phoenician god

2. **causeway** — a built-up road

3. **Yehovah** — a Hebrew name for God, usually spelled Jehovah, but there is no "J" in the Hebrew alphabet

4. **scroll** — papyrus or animal skin document rolled up into a cylinder

5. **annunciation** — an announcement

6. **Calvary** — Latin word meaning "skull"

7. **Golgotha** — Hebrew word meaning "skull"

8. **grotto** — cave

9. **Messiah** — meaning "Anointed One" and applied to an expected Jewish leader

10. **ossuary** — a box in which human bones were preserved

11. Gebal, Berytus Sidon and Tyre

12. Cyprus

13. Ahiram

14. Elijah

15. Ezekiel

16. 1947

17. Qumram

18. Constantine

19.

20. a. Psalms

b. Lamentations

Archaeology Book - Test Answer Key

1. **carbon dating** — calculating the amount of carbon left in organic material that has died
2. **baulk** — the vertical ridge left between two excavated squares in the ground
3. **synchronism** — something happening at the same time
4. **mastabas** — mud-brick structures beneath which were tomb chambers
5. **centurion** — a military officer in charge of a hundred men
6. **Chaldees** — people who used to live in southern Iraq
7. **bulla** — an impression made on clay with a seal (plural: bullae)
8. **cuneiform** — a form of writing using a wedge-shaped stylus to make an impression on a clay tablet
9. **syncline** — a boat-shaped geological formation
10. **Persia** — a country in central Iran
11. **rhyton** — a drinking vessel shaped like a human or animal
12. **annunciation** — an announcement
13. **ossuary** — a box in which human bones were preserved
14. **grotto** — cave
15. Early Bronze, Middle Bronze, Late Bronze, Iron Age
16. Any two: defense, heat, and floods
17. Gebal, Berytus Sidon and Tyre
18. Zoser
19. Khufu
20. Jehu
21. Hezekiah
22. Isaiah
23. Cyrus the Great
24. Ezekiel
25. 1947
26. Identify the Pyramids, Temples, Tombs, and unique features on Giza Map:

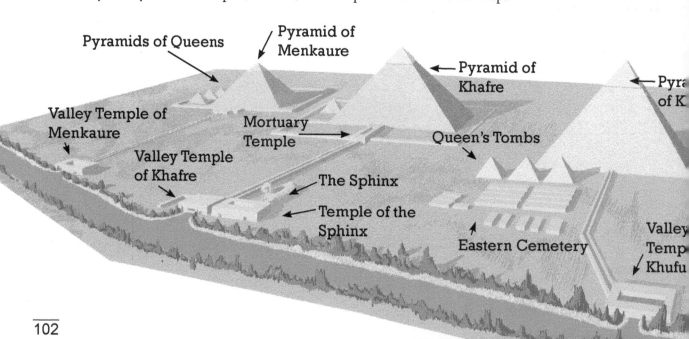

27.

Hebrew

Sumerian

Phoenician

Egyptian

27. a. Psalms

b. Lamentations

Geology - Unit Quiz Answer Key

Unit One Quiz, chapters 1-3

1. **Principle of uniformity** — the scientific thought that past proccesses are no different than processes today, meaning everything happens by gradual process over very long periods of time

2. **Principle of catastrophe** — the scientific thought that sees evidence of rapid, highly energetic events over short periods of time, doing a lot of geologic work

3. **Sediment** — natural materials broken down by processes of erosion and weathering; can be transported or deposited by water or wind

4. **Metamorphism** — a process of heat and pressure that causes one rock to alter into another

5. Uniformity (the present is the key to the past) and catastrophe (highly energetic events operated over short periods of time and did much geologic work rapidly)

6. Day 1: Earth, space, time, light; Day 2: Atmosphere; Day 3: dry land, plants; Day 4: sun, moon, stars, planets; Day 5: sea and flying creatures; Day 6: land animals, people

7. Crust, mantle, outer core, inner core

8. Fault: rock is broken and shoved up (Colorado Plateau); warped: regional squeezing or slow uplift (Appalachians); lava: hardened lava plains that may have been uplifted or hardened at the current level (Columbia River basalts)

9. Folded: layers of sediments that have been crumpled by pressures from the side (Alps, Himalayas, Appalachians, Rocky Mountains); domed: sediments pushed up from below (Black Hills of South Dakota); fault block: one area of sediments are pushed up (Grand Teton Mountains); volcanic: molten lavas pushed out to the surface of the earth (Hawaii's volcanic islands, Mount Rainier, Mount St. Helens, Mount Ararat)

10. a. Granite: Igneous, b. Marble: Metamorphic, c. Shale: Classic Sedimentary, d. Limestone: Organic Chemical Sedentary, e. Coal: Organic Chemical Sedentary, f. Rhyolite: Igneous, g. Slate: Metamorphic

Unit Two Quiz, chapter 4

1. **Erosion** — states if processes continue as they do today, everything will eventually be eroded or worn away

2. **Petrification** — the process by which trees, plants, and even animals are solidified by burial in hot, silica-rich water

3. **Turbidite** — a underwater, rapid deposition of mud which hardens into a layer of rock

4. **Gastrolith** — rounded stones which were used by plant-eating dinosaurs to aid in digestion and sometimes found with fossilized remains

5. **Fumaroles** — an opening in the earth's crust, usually associated with volcanic activity

6. **Carbon dating** — a process which uses the decay of carbon 14 to estimate the age of things that were once living

7. Rain, ice, plants and animals, chemicals, ocean waves

8. Any four: hard parts are preserved; replacement by other minerals; cast or mold are all that remains; petrification; cabonization; preservation of soft parts; frozen animals; animal tracks and worm burrows; coprolites; gastroliths

9. Radioactive; uranium

10. Silica

11. First, an event such as an earthquake starts a mud flow underwater. Next, the mud flow spreads out. Eventually the mud flow hardens into a layer of rock.

12. The organism must be buried rapidly, protected from scavengers and from decomposition by bacteria and chemicals.

13. Whether it is soft or brittle, how deep it is buried

14. First, layers of sediment are deposited. The weight of the water and the sediments on top begin to compact the sediments underneath. Next, warm water circulates throughout the sediments and dissolves certain minerals. The dissolved minerals surround the individual grains of sediment. Finally, when the water cools off and stops moving, the dissolved minerals act as a "glue" that cements the grains of sediment together to form sedimentary rock.

15. Heat and pressure recrystallize the minerals in rock into new mineral combinations. Some believe it happened over long periods of time; others believe it happened over short periods of time.

16. When a plant is living, it takes the isotope carbon 14 into its leaves, stems, and seeds. After the plant dies, the carbon 14 decays into nitrogen 14. Scientists can measure the amounts of both carbon 14 and carbon 12. Since they know the time it takes the isotope to decay, they can calculate when the plant died.

17. Answers will vary but should include key ideas: First, if a dinosaur was not on the ark, then it drowned in the flood. Next, the animal was buried rapidly as the flood deposited soft layers of material that later hardened into stone. Then, a process of fossilization occurred, such as the bones being replaced by dissolved minerals in the ground water. Finally, the fossils became exposed as the ground around the animal eroded away.

Unit Three Quiz, chapters 5-6

1. **Magnetic field** — A field that exerts forces on objects made of magnetic materials; made up of many lines of force.

2. **Uplift** — in geology, a tectonic uplift is a geological process most often caused by plate tectonics which causes an increase in elevation

3. **Second law of science** — also referred to as the second law of thermodynamics, which states that in every process or reaction in the universe the components deteriorate

4. **Fountains of the deep** — a phrase mentioned in Genesis 7 as a reference to sources of water as part of the Great Flood of Noah.

5. **Glacier** — a huge mass of ice that moves slowly over land

6. **Polar ice cap** — a high latitude region of a planet that is covered in ice

7. Creation, the Fall, Flood, Ice Age

8. God sent the Flood as a judgment on the wickedness of mankind. The Flood formed many of the rockand fossil layers. See pages 63-66 for more details.

9. a. chemicals in the ocean, b. erosion of the continents, c. sediments in the ocean, d. dating the atmosphere, e. dating the magnetic field

10. The Bible says in Genesis 3 that the entire creation came under the curse of sin, including plants, animals, mankind, and the earth. As a result of the curse, everything is wearing down and deteriorating.

11. A majority of methods used to age-date the earth yield ages far less than the acclaimed billions of years.

12. Formed the cores of the continents, some erosion and deposition probably happened

13. The top of Mount Everest was once underwater and was later pushed up after the Flood waters receded.

14. The warm ocean waters rapidly evaporated and condensed over the colder continents, causing a buildup of ice and snow. See page 67 for more detailed information.

15. Review answer against student's previous essay on this subject.

Unit Four Quiz, chapters 7-8

1. **Volcanism** — the eruption of molten rock (magma) onto the surface of the earth

2. **Escarpment** — a steep slope or long cliff that occurs from erosion or faulting and separates two relatively level areas of differing elevations

3. At the end of Noah's flood it appears that a great volume of water was trapped, held in place by the Kaibab Upwarp. Ice Age rains filled the lake to overflowing and as it burst through its mountain "dam," the huge volume of lake waters carved the canyon.

4. At Yellowstone Park, the soil and rock is thin, allowing very hot material to be near the surface. As rain and run-off water trickle down into the earth they get heated, bubbling up in places as hot springs. In some places the underground water is trapped and when heated to an excessive degree, it bursts out in a geyser.

5. The level of water in Lake Erie is somewhat higher than the elevation of nearby Lake Ontario. A river draining the waters of Lake Erie into Lake Ontario runs over the Niagara escarpment, resulting in a spectacular set of falls. Erosion takes place as the water roars over the falls, and the escarpment naturally recedes toward Lake Erie.

6. Both mountain chains are the result of layers of sediments deposited by Noah's flood. The Appalachians buckled up in the early stages of the flood and were subjected to massive erosion by the continuing flood waters. The Rocky Mountains buckled up late in the flood, extending up above the flood waters as the waters drained off. Thus, the erosion to which they were subjected was much less intense.

7. Petrified wood can form, under laboratory conditions, in a very short period of time. The speed of petrifaction is related to the pressures which inject the hot silica-rich waters into the wood.

8. When water saturated with calcium carbonate enters an open space such as a cave, it cools off or evaporates, leaving the calcium carbonate behind. Stalactites (holding "tightly" to the ceiling) and stalagmites (which are usually larger and thus more "mighty" than stalactites) are formed as this calcium carbonate precipitates out of the water.

9. Coal can be formed in a laboratory by heating organic material away from oxygen but in the presence

of volcanic clay. Under such conditions, coal can be formed in a matter of hours. One wonders if the abundant forest growing before the flood would not have formed huge log mats floating on the flood ocean. As these decayed and were buried by hot sediments in the presence of volcanic clay, they might have rapidly turned to coal.

10. Natural gas, mostly methane, is given off in the coalification process. The largest quantities of it, however, are found in deep rocks not associated with decomposition of organic material. Evidently, some natural gas is from both organic and inorganic sources.

11. Many theories have been promoted as to the specific origin of oil. The best seems to be that it is the remains of algae once floating in the ocean but buried in ocean sediments. Oil is not the remains of dinosaurs as has sometimes been claimed.

12. No - they are rare. Most fossils are of sea creatures, fish, and insects. Only relatively few fossils are of land animals, specifically dinosaurs.

13. Answers will vary.

14. Answers will vary.

Geology Book - Test Answer Key

1. **Principle of uniformity** — the scientific thought that past proccesses are no different than processes today, meaning everything happens by gradual process over very long periods of time

2. **Principle of catastrophe** — the scientific thought that sees evidence of rapid, highly energetic events over short periods of time, doing a lot of geologic work

3. **Erosion** — the process by which soil and rock are worn away

4. **Petrification** — the process by which trees, plants, and even animals are solidified by burial in hot, silica-rich water

5. **Turbidite** — a underwater, rapid deposition of mud which hardens into a layer of rock

6. **Gastrolith** — rounded stones which were used by plant-eating dinosaurs to aid in digestion and sometimes found with fossilized remains

7. **Fumaroles** — an opening in the earth's crust, usually associated with volcanic activity

8. **Metamorphism** — a process of heat and pressure that causes one rock to alter into another

9. **Magnetic field** — a field that exerts forces on objects made of magnetic materials; made up of many lines of force

10. **Sediment** — natural materials broken down by processes of erosion and weathering; can be transported or deposited by water or wind

11. **Second law of science** — also referred to as the second law of thermodynamics, which states that in every process or reaction in the universe the components deteriorate

12. **Fountains of the deep** — a phrase mentioned in Genesis 7 as a reference to sources of water as part of the Great Flood of Noah.

13. **Glacier** — a huge mass of ice that moves slowly over land

14. **Volcanism** — the eruption of molten rock (magma) onto the surface of the earth

15. Uniformity (the present is the key to the past) and catastrophe (highly energetic events operated over short periods of time and did much geologic work rapidly)

16. Crust, mantle, outer core, inner core

17. Creation, the Fall, Flood, Ice Age

18. The organism must be buried rapidly, protected from scavengers and from decomposition by bacteria and chemicals.

19. Whether it is soft or brittle, how deep it is buried

20. A majority of methods used to age-date the earth yield ages far less than the acclaimed billions of years.

21. Formed the cores of the continents, some erosion and deposition probably happened

22. The top of Mount Everest was once underwater and was later pushed up after the Flood waters receded.

23. The warm ocean waters rapidly evaporated and condensed over the colder continents, causing a buildup of ice and snow. See page 67 for more detailed information.

24. God sent the Flood as a judgment on the wickedness of mankind.

25. At the end of Noah's flood it appears that a great volume of water was trapped, held in place by the Kaibab Upwarp. Ice Age rains filled the lake to overflowing and as it burst through its mountain "dam," the huge volume of lake waters carved the canyon.

26. The level of water in Lake Erie is somewhat higher than the elevation of nearby Lake Ontario. A river draining the waters of Lake Erie into Lake Ontario runs over the Niagara escarpment, resulting in a spectacular set of falls. Erosion takes place as the water roars over the falls, and the escarpment naturally recedes toward Lake Erie.

27. Both mountain chains are the result of layers of sediments deposited by Noah's flood. The Appalachians buckled up in the early stages of the flood and were subjected to massive erosion by the continuing flood waters. The Rocky Mountains buckled up late in the flood, extending up above the flood waters as the waters drained off. Thus, the erosion to which they were subjected was much less intense.

28. Petrified wood can form, under laboratory conditions, in a very short period of time. The speed of petrifaction is related to the pressures which inject the hot silica-rich waters into the wood.

29. a. Granite: Igneous, b. Marble: Metamorphic, c. Shale: Classic Sedimentary, d. Limestone: Organic Chemical Sedentary, e. Coal: Organic Chemical Sedentary, f. Rhyolite: Igneous, g. Slate: Metamorphic

Archaeology Book - Glossary

Accession year—the year a king actually began his reign

AD—Anno Domini (the year of our Lord)—the years after the Christian era began

Amphitheater—a circle of seats surrounding an area where gladiators fought each other or fought wild beasts

Anatolia—mountainous area in central Turkey

Annunciation—an announcement

Armenians—people who live in eastern Turkey and northern Iraq

Artifact—an item from antiquity found in excavation

Asiatic—in Egyptian terms, someone from Syria or Palestine

Baal—a word meaning "lord." The name of a Phoenician god

Bathhouse—a club where citizens could bathe in cold, warm, or hot water

BC—Before Christ; the years before the Christian era began

Baulk—the vertical ridge left between two excavated squares in the ground

Bedouin—Arabs living in tents with no fixed address

Bulla—an impression made on clay with a seal (plural: bullae)

Calvary—Latin word meaning "skull"

Carbon dating—calculating the amount of carbon left in organic material that has died

Causeway—a built-up road

Centurion—a military officer in charge of a hundred men

Ceramic—something made of pottery

Chaldees—people who used to live in southern Iraq

Chronology—time periods; dates in which events happened

Cistern—a hole dug in rock to store rainwater

Cuneiform—a form of writing using a wedge-shaped stylus to make an impression on a clay tablet

Debris—discarded rubbish

Dowry—gift given to a prospective bride at the time of her marriage

Dynasty—a succession of kings descended from one another

Drachma—a Greek coin worth about a day's wages

EB—the Early Bronze Period

Edom—country in southern Jordan

Edomites—people descended from Edom, also known as Esau, Jacob's brother

Exile—a people sent out of their home country to another country

Exodus—going out; applied to the Israelites leaving Egypt

Golgotha—Hebrew word meaning "skull"

Grotto—cave

Hieroglyphs—Egyptian picture writing

Inscription—writing made on clay, stone, papyrus, or animal skins

LB—the Late Bronze period

Mastabas—mud-brick structures beneath which were tomb chambers

MB—the Middle Bronze Period

Medes—people who used to live in northern Iran

Messiah—meaning "Anointed One" and applied to an expected Jewish leader

Millennium—one thousand years

Nabataeans—people descended from Nabaioth who occupied Petra

Nomad—a person who lived in a tent that could be moved from place to place

Non-accession year—the first complete year of a king's reign

Nubia—a country south of Egypt now called Sudan

Ossuary—a box in which human bones were preserved

Papyrus—sheets of writing material made from the Egyptian papyrus plant

Passover—Jewish ceremony celebrating the Exodus from Egypt

Persia—a country in central Iran

Pharaoh—title applied to many Egyptian kings

Pottery—a vessel made of clay fired in a kiln

Rhyton—a drinking vessel shaped like a human or animal

Sarcophagus—a stone coffin

Scarab—model of a dung beetle with an inscription engraved on it for sealing documents

Seal—an object made of stone, metal, or clay with a name engraved on it used to impress in soft clay

Scroll—papyrus or animal skin document rolled up into a cylinder

Siq—narrow valley between two high rock formations

Stratum—a layer of occupation exposed by excavations (plural: strata)

Synchronism—something happening at the same time

Syncline—a boat-shaped geological formation

Tell—a Hebrew word meaning "ruins" applied to hills on which people once lived

Theater—a stage for actors in front of which was a semi-circle of seats

Vellum—animal skin treated to be used as writing material

Wadi—a dry riverbed, carrying water only when it rained

Yehovah—a Hebrew name for God, usually spelled Jehovah, but there is no "J" in the Hebrew alphabet

Geology Book - Glossary

Alluvial Sediment—material carried by fast-moving rivers and streams that are deposited at points where the water moves slower

Asthenosphere—a suspected area in the uppermost portion of the earth's mantle where material is hot and deforms easily

Atom—the basic component of chemical elements

Basalt—a type of igneous rock that makes up most of the oceanic crust; on land it forms when extruded by volcanoes or through fissures

Canyon—large areas with steep walls that have been carved out of layers of sedimentary rock

Carbon dating—a process which uses the decay of carbon 14 to estimate the age of things that were once living

Catastrophism—the philosophy about the past that allows for totally different processes and/or rates, scales, and intensities than those operating today; includes the idea that processes such as creation and dynamic global flooding have shaped the entire planet

Cavitation—bubbles within fast moving water explode inwardly and pound against a rock, eventually reducing to powder

Cementation—a process in the formation of sedimentary rock when minerals are dissolved which then help to solidify the rock

Chemical rock—a type of sedimentary rock built up as chemicals in water, usually seawater, precipitate and consolidate

Clastic rock—a type of sedimentary rock consisting of fragments of a previously existing rock (i.e., sandstone consists of consolidated sand grains)

Concretions—concreted masses of sedimentary rock that has been eroded out of a softer area of rock

Continental separation—the concept that the continental plates have moved apart (or collided), concluding, for example, that Africa and South America were once connected

Continental Shield—the primarily granite core of a continent that has been exposed to the surface and then bulges up because there is no weight upon it

Compaction—a process in the formation of sedimentary rock when the materials are pushed together tightly, leaving little to no open spaces

Coprolite—fossilized animal or dinosaur dung; can be used to determine a creature's diet

Core—the center of the earth is thought to be a sphere of iron and nickel, divided into two zones. The outer core is in molten or liquid form, while the inner core is solid.

Crossbed—areas of extremely large ripple marks

Crust—the thin covering of planet Earth, which includes the continents and ocean basins. Nowhere is it more than 60 miles (100 km) thick.

Decomposition—the process by which things are broken down into smaller or more basic substances or elements

Deposition—the process by which sediments are deposited onto a landform

Earthquake—a sudden release of energy below the earth's crust which cause the earth's crust to move or shake

Erosion—the process by which soil and rock are worn away

Escarpment—a steep slope or long cliff that occurs from erosion or faulting and separates two relatively level areas of differing elevations

Fault—a fracture in rock along which separation or movement has taken place

Fold—a bend or flexure in a layer of rock

Fossil—the direct or indirect remains of an animal or plant

Fountains of the deep—a phrase mentioned in Genesis 7 as a reference to sources of water as part of the Great Flood of Noah. Some biblical scholars feel it could refer to oceanic or subterranean sources of water.

Fumaroles—an opening in the earth's crust, usually associated with volcanic activity

Gastrolith—rounded stones which were used by plant-eating dinosaurs to aid in digestion and sometimes found with fossilized remains

Geyser—underground water that has been heated to an excessive degree and because of pressure bursts out of the ground temporarily

Glacier—a large natural formation of ice where the accumulation of ice and snow exceeds the amount it melts or turns from a solid to a gas

Granite—a widespread igneous rock, which contains abundant quartz and feldspar and makes up a significant portion of the continental crust

Igneous rock—rock formed when hot, molten magma cools and solidifies

Isotope—variations of an element's atoms, usually in the different number of neutrons

Kolk—an underwater tornado which lifts or removes underlying materials

Law of disintegration—states if processes continue as they do today, everything will eventually be eroded or worn away

Magnetic field—a field that exerts forces on objects made of magnetic materials; made up of many lines of force

Mantle—beneath the thin crust and above the core of the earth. It is about 1,864 miles (3,000 km) thick

Metamorphism—a process of heat and pressure that causes one rock to alter into another

Metamorphic rock—rocks formed when heat, pressure, and/or chemical action alters previously existing rock

Mountain—a large landform rising abruptly from the surrounding area

Obsidian—a common type of rhyolitic volcanic rock, which almost looks like a chunk of black glass

Petrification—the process by which trees, plants, and even animals are solidified by burial in hot, silica-rich water

Plain—a broad area of relatively flat land

Plate—the earth's crust, both continental and oceanic, is divided into plates, with boundaries identified by zones of earthquake activity. The idea of plate tectonics holds that these plates move relative to one another, sometimes separating or colliding, and sometimes moving past each other.

Plateau—flat lying sediment layers similar to plains but at higher elevations

Polar ice cap—a high latitude region of a planet that is covered in ice

Principle of uniformity—the scientific thought that past proccesses are no different than processes today, meaning everything happens by gradual process over very long periods of time

Principle of catastrophe—the scientific thought that sees evidence of rapid, highly energetic events over short periods of time, doing a lot of geologic work

Radioisotope Dating—the attempt to determine a rock's age by measuring the ratio of radioactive isotopes and the rate at which they decay

Rhyolite—molten rock which forms granite that has erupted on land and solidified

Ripple Marks—marks which indicate moving water flowed over a rock layer when the sediments were still muddy and yet to harden

Second law of science—also referred to as the second law of thermodynamics, which states that the entropy (disorder) of the Universe increases over time

Sediment—natural materials broken down by processes of erosion and weathering; can be transported or deposited by water or wind

Sedimentary rock—rock formed by the deposition and consolidation of loose particles of sediment, and those formed by precipitation from water

Sedimentation—the act of depositing sediments

Tsunami—often called a tidal wave; a seismic sea wave produced by an underwater disturbance such as an earthquake, volcano or landslide. These can be extremely destructive.

Turbidite—a underwater, rapid deposition of mud which hardens into a layer of rock

Uniformitarianism—the philosophy about the past which assumes no past events of a different nature than those possible today, and/or operating at rates, scales and intensities far greater than those operating today. The slogan "the present is the key to the past" characterizes this idea

Uplift—in geology, a tectonic uplift is a geological process most often caused by plate tectonics which causes an increase in elevation

Volcanism—this is the process by which molten rock or lava erupts through the surface of a planet

Parent Lesson Plans

Now turn your favorite Master Books into curriculum! Each complete three-hole punched Parent Lesson Plan (PLP) includes:

- An easy-to-follow, one-year educational calendar
- Helpful worksheets, quizzes, tests, and answer keys
- Additional teaching helps and insights
- Complete with all you need to quickly and easily begin your education program today!

ELEMENTARY ZOOLOGY

1 year
4th – 6th

Package Includes: *World of Animals, Dinosaur Activity Book, The Complete Aquarium Adventure, The Complete Zoo Adventure, Parent Lesson Plan*

5 Book Package $80.99
978-0-89051-747-5

SCIENCE STARTERS: ELEMENTARY PHYSICAL & EARTH SCIENCE

1 year
3rd – 8th grade

6 Book Package Includes: *Forces & Motion –Student, Student Journal, and Teacher; The Earth – Student, Teacher & Student Journal; Parent Lesson Plan*

6 Book Package $51.99
978-0-89051-748-2

SCIENCE STARTERS: ELEMENTARY CHEMISTRY & PHYSICS

1 year
3rd – 8th grade

Package Includes: *Matter – Student, Student Journal, and Teacher; Energy – Student, Teacher, & Student Journal; Parent Lesson Plan*

7 Book Package $54.99
978-0-89051-749-9

INTRO TO METEOROLOGY & ASTRONOMY

1 year
7th – 9th grade
½ Credit

Package Includes: *The Weather Book; The Astronomy Book; Parent Lesson Plan*

3 Book Package $44.99
978-0-89051-753-6

INTRO TO OCEANOGRAPHY & ECOLOGY

1 year
7th – 9th grade
½ Credit

Package Includes: *The Ocean Book; The Ecology Book; Parent Lesson Plan*

3 Book Package $45.99
978-0-89051-754-3

INTRO TO SPELEOLOGY & PALEONTOLOGY

1 year
7th – 9th grade
½ Credit

Package Includes: *The Cave Book; The Fossil Book; Parent Lesson Plan*

3 Book Package $44.99
978-0-89051-752-9

CONCEPTS OF MEDICINE & BIOLOGY

1 year
7th – 9th grade
½ Credit

Package Includes: *Exploring the History of Medicine; Exploring the World of Biology; Parent Lesson Plan*

3 Book Package $41.99
978-0-89051-756-7

CONCEPTS OF MATHEMATICS & PHYSICS

1 year
7th – 9th grade
½ Credit

Package Includes: *Exploring the World of Mathematics; Exploring the World of Physics; Parent Lesson Plan*

3 Book Package $41.99
978-0-89051-757-4

CONCEPTS OF EARTH SCIENCE & CHEMISTRY

1 year
7th – 9th grade
½ Credit

Package Includes: *Exploring Planet Earth; Exploring the World of Chemistry; Parent Lesson Plan*

3 Book Package $41.99
978-0-89051-755-0

THE SCIENCE OF LIFE: BIOLOGY

1 year
8th – 9th grade
½ Credit

Package Includes: *Building Blocks in Science; Building Blocks in Life Science; Parent Lesson Plan*

3 Book Package $45.99
978-0-89051-758-1

BASIC PRE-MED

1 year
8th – 9th grade
½ Credit

Package Includes: *The Genesis of Germs; The Building Blocks in Life Science; Parent Lesson Plan*

3 Book Package $43.99
978-0-89051-759-8

Parent Lesson Plans

INTRO TO ASTRONOMY

1 year
7th – 9th grade
½ Credit

Package Includes: *The Stargazer's Guide to the Night Sky; Parent Lesson Plan*

2 Book Package $45.99
978-0-89051-760-4

INTRO TO ARCHAEOLOGY & GEOLOGY

1 year
7th – 9th
½ Credit

Package Includes: *The Archaeology Book; The Geology Book; Parent Lesson Plan*

3 Book Package $45.99
978-0-89051-751-2

SURVEY OF SCIENCE HISTORY & CONCEPTS

1 year
10th – 12th grade
1 Credit

Package Includes: *The World of Mathematics; The World of Physics; The World of Biology; The World of Chemistry; Parent Lesson Plan*

5 Book Package $72.99
978-0-89051-764-2

SURVEY OF SCIENCE SPECIALTIES

1 year
10th – 12th grade
1 Credit

Package Includes: *The Cave Book; The Fossil Book; The Geology Book; The Archaeology Book; Parent Lesson Plan*

5 Book Package $81.99
978-0-89051-765-9

SURVEY OF ASTRONOMY

1 year
10th – 12th grade
1 Credit

Package Includes: *The Stargazers Guide to the Night Sky; Our Created Moon; Taking Back Astronomy; Our Created Moon DVD; Created Cosmos DVD; Parent Lesson Plan*

4 Book, 2 DVD Package $108.99
978-0-89051-766-6

GEOLOGY & BIBLICAL HISTORY

1 year
8th – 9th
1 Credit

Package Includes: *Explore the Grand Canyon; Explore Yellowstone; Explore Yosemite & Zion National Parks; Your Guide to the Grand Canyon; Your Guide to Yellowstone; Your Guide to Zion & Bryce Canyon National Parks; Parent Lesson Plan.*

4 Book, 3 DVD Package $112.99
978-0-89051-750-5

PALEONTOLOGY: LIVING FOSSILS

1 year
10th – 12th grade
½ Credit

Package Includes: *Living Fossils, Living Fossils Teacher Guide, Living Fossils DVD; Parent Lesson Plan*

3 Book, 1 DVD Package $66.99
978-0-89051-763-5

LIFE SCIENCE ORIGINS & SCIENTIFIC THEORY

1 year
10th – 12th grade
1 Credit

Package Includes: *Evolution: the Grand Experiment, Teacher Guide, DVD; Living Fossils, Teacher Guide, DVD; ParentLesson Plan*

5 Book, 2 DVD Package $133.99
978-0-89051-761-1

NATURAL SCIENCE THE STORY OF ORIGINS

1 year
10th – 12th grade
½ Credit

Package Includes: *Evolutions the Grand Experiment; Evolution the Grand Experiment Teacher's Guide, Evolution the Grand Experiment DVD; Parent Lesson Plan*

3 Book, 1 DVD Package $66.99
978-0-89051-762-8

ADVANCED PRE-MED STUDIES

1 year
10th – 12th grade
1 Credit

Package Includes: *Building Blocks in Life Science; The Genesis of Germs; Body by Design; Exploring the History of Medicine; Parent Lesson Plan*

5 Book Package $78.99
978-0-89051-767-3

BIBLICAL ARCHAEOLOGY

1 year
10th – 12th grade
1 Credit

Package Includes: *Unwrapping the Pharaohs; Unveiling the Kings of Israel; The Archaeology Book; Parent Lesson Plan.*

4 Book Package $101.99
978-0-89051-768-0

CHRISTIAN HERITAGE

1 year
10th – 12th grade
1 Credit

Package Includes: *For You They Signed; Lesson Parent Plan*

2 Book Package $45.99
978-0-89051-769-7